797,885 Books
are available to read at

Forgotten Books

www.ForgottenBooks.com

Forgotten Books' App
Available for mobile, tablet & eReader

ISBN 978-1-332-35233-3
PIBN 10317654

This book is a reproduction of an important historical work. Forgotten Books uses state-of-the-art technology to digitally reconstruct the work, preserving the original format whilst repairing imperfections present in the aged copy. In rare cases, an imperfection in the original, such as a blemish or missing page, may be replicated in our edition. We do, however, repair the vast majority of imperfections successfully; any imperfections that remain are intentionally left to preserve the state of such historical works.

Forgotten Books is a registered trademark of FB &c Ltd.
Copyright © 2015 FB &c Ltd.
FB&c Ltd, Dalton House, 60 Windsor Avenue, London, SW19 2RR.
Company number 08720141. Registered in England and Wales.

For support please visit www.forgottenbooks.com

1 MONTH OF FREE READING

at
www.ForgottenBooks.com

By purchasing this book you are eligible for one month membership to ForgottenBooks.com, giving you unlimited access to our entire collection of over 700,000 titles via our web site and mobile apps.

To claim your free month visit: www.forgottenbooks.com/free317654

* Offer is valid for 45 days from date of purchase. Terms and conditions apply.

English
Français
Deutsche
Italiano
Español
Português

www.forgottenbooks.com

Mythology Photography **Fiction**
Fishing Christianity **Art** Cooking
Essays Buddhism Freemasonry
Medicine **Biology** Music **Ancient Egypt** Evolution Carpentry Physics
Dance Geology **Mathematics** Fitness
Shakespeare **Folklore** Yoga Marketing
Confidence Immortality Biographies
Poetry **Psychology** Witchcraft
Electronics Chemistry History **Law**
Accounting **Philosophy** Anthropology
Alchemy Drama Quantum Mechanics
Atheism Sexual Health **Ancient History**
Entrepreneurship Languages Sport
Paleontology Needlework Islam
Metaphysics Investment Archaeology
Parenting Statistics Criminology
Motivational

REPORT

UPON

WEIGHTS AND MEASURES,

BY

JOHN QUINCY ADAMS,

SECRETARY OF STATE OF THE UNITED STATES.

PREPARED IN OBEDIENCE TO A RESOLUTION OF THE SENATE OF THE THIRD MARCH, 1817.

WASHINGTON:

PRINTED BY GALES & SEATON

1821.

DEPARTMENT OF STATE,

February 22, 1821.

SIR: I have the honor of transmitting, herewith, a Report upon Weights and Measures, prepared in conformity to a resolution of the Senate of the 3d March, 1817.

With the highest respect,
I am, Sir,
Your very humble and ob't servant,
JOHN QUINCY ADAMS.

*To the President of the Senate
of the United States.*

REPORT.

The Secretary of State, to whom, by a resolution of the Senate of the 3d of March, 1817, it was referred to prepare and report to the Senate, "a statement relative to the regulations and standards for weights and measures in the several states, and relative to proceedings in foreign countries, for establishing uniformity in weights and measures, together with such propositions relative thereto, as may be proper to be adopted in the United States," respectfully submits to the Senate the following

REPORT:

The resolution of the Senate embraces three distinct objects of attention, which it is proposed to consider in the following order:

1. The proceedings in foreign countries for establishing uniformity in weights and measures.
2. The regulations and standards for weights and measures in the several states of the Union.
3. Such propositions relative to the uniformity of weights and measures as may be proper to be adopted in the United States.

The term *uniformity*, as applied to weights and measures, is susceptible of various constructions and modifications, some of which would restrict, while others would enlarge, the objects in contemplation by the resolution of the Senate.

Uniformity in weights and measures may have reference

1. To the weights and measures themselves.
2. To the objects of admeasurement and weight.
3. To *time*, or the duration of their establishment.
4. To *place*, or the extent of country over which, including the persons by whom, they are used.
5. To numbers, or the modes of numeration, multiplication, and division, of their parts and units.
6. To t ei *nomenclature*, or the denominations by which they are calledr
7. To their connection with *coins* and *moneys of account*.

In reference to the weights and measures themselves, there may be
    An uniformity of identity, or
    An uniformity of proportion.

By an uniformity of *identity*, is meant a system founded on the principle of applying only one unit of weights to all weighable articles, and one unit of measures of capacity to all substances, thus measured, liquid or dry.

By an uniformity of *proportion*, is understood a system admitting more than one unit of weights, and more than one of measures of capacity; but in which all the weights and measures of capacity are in a uniform proportion with one another.

Our present existing weights and measures are, or originally were, founded upon the uniformity of proportion. The new French metrology is founded upon the uniformity of identity.

And, in reference to each of these circumstances, and to each in combination with all, or either of the others, uniformity may be more or less extensive, partial, or complete.

Measures and weights are the instruments used by man for the comparison of quantities, and proportions of things.

In the order of human existence upon earth, the objects which successively present themselves, are man—natural, domestic, civil society, government, and law. The want, at least, of measures of length, is founded in the physical organization of individual man, and precedes the institution of society. Were there but one man upon earth, a solitary savage, ranging the forests, and supporting his existence by a continual conflict with the wants of his nature, and the rigor of the elements, the necessities for which he would be called to provide would be *food, raiment, shelter*. To provide for the wants of food and raiment, the first occupation of his life would be the chase of those animals, the flesh of which serves him for food, and the skins of which are adaptable to his person for raiment. In adapting the raiment to his body, he would find at once, in his own person, the want and the supply of a standard measure of length, and of the proportions and subdivisions of that standard.

But, to the continued existence of the human species, two persons of different sexes are required. Their union constitutes natural society, and their permanent cohabitation, by mutual consent, forms the origin of domestic society. Permanent cohabitation requires a common place of abode, and leads to the construction of edifices where the associated parties, and their progeny, may abide. To the construction of a dwelling place, superficial measure becomes essential, and the dimensions of the building still bear a natural proportion to those of its destined inhabitants. Vessels of capacity are soon found indispensable for the supply of water; and the range of excursion around the dwelling could scarcely fail to suggest the use of a measure of itinerary distance.

Measures of *length*, therefore, are the wants of individual man, independent of, and preceding, the existence of society. Measures of surface, of distance, and of capacity, arise immediately from domestic society. They are wants proceeding rather from social, than from individual, existence. With regard to the first, *linear* measure, nature in creating the want, and in furnishing to man, within himself, the means of its supply, has established a system of numbers, and of proportions, between the man, the measure, and the objects measured. Linear measure requires only a change of direction to become a measure of circumference; but is not thereby, without calculation, a measure of surface. Itinerary measure, as it needs nothing more than the prolongation or repetition of linear measure, would seem at the first view to be the same. Yet this is evidently not the progress of nature. As the want of it originates in a different stage of human existence, it will not naturally occur to man, to use the same measure, or the same scale of proportions and numbers, to clothe his body, and to mark the distance of his walks. On the contrary, for the measurement of all objects which he can lift and handle, the fathom, the arm, the cubit, the hand's-breadth, the span, and the fingers, are the instruments proposed to him by nature; while the pace and the foot are those which she gives him for the measurement of itinerary distance. These natural standards are never, in any stage of society, lost to individual man. There are probably few persons living who do not occasionally use their own arms, hands, and fingers, to measure objects which they handle, and their own pace to measure a distance upon the ground.

Here then is a source of *diversity*, to the standards even of linear measure, flowing from the difference of the relations between man and physical nature. It would be as inconvenient and unnatural to the organization of the human body to measure a bow and arrow for instance, the first furniture of solitary man, by his foot or pace, as to measure the distance of a day's journey, or a morning's walk to the hunting ground, by his arm or hand.

Measures of capacity are rendered necessary by the nature of fluids, which can be held together in definite quantities only by vessels of substance more compact than their own. They are also necessary for the admeasurement of those substances which nature produces in multitudes too great for numeration, and too minute for linear measure. Of this character are all the grains and seeds, which, from the time when man becomes a tiller of the ground, furnish the principal materials of his subsistence. But nature has not furnished him with the means of supplying this want in his own person. For this measure he is obliged to look abroad into the nature of things; and his first measure of capacity will most probably be found in the egg of a large bird, the shell of a cetaceous fish, or the horn of a beast. The want of a *common* standard not being yet felt, these measures will be of various dimensions; nor is it to be expected that the thought will ever occur to the man of nature, of establishing a proportion between his cubit and his cup, of graduating his pitcher by the size of his foot, or equalizing its parts by the number of his fingers.

Measures of length, once acquired, may be, and naturally are, applied to the admeasurement of objects of surface and solidity; and hence arise new diversities from the nature of things. The connection of linear measure with *numbers*, necessarily, and in the first instance, imports only the first arithmetical rule of numeration, or addition. The mensuration of surfaces, and of solids, requires the further aid of multiplication and division. Mere numbers, and mere linear measure, may be reckoned by addition alone; but their application to the surface can be computed only by multiplication. The elementary principle of decimal arithmetic is then supplied by nature to man within himself in the number of his fingers. Whatever standard of linear measure he may assume, in order to measure the surface or the solid, it will be natural to him to stop in the process of addition when he has counted the tale equal to that of his fingers. Then turning his line in the other direction, and stopping at the same term, he finds the square of his number a hundred: and, applying it again to the solid, he finds its cube a thousand.

But while decimal arithmetic thus, for the purposes of *computation*, shoots spontaneously from the nature of man and of things, it is not equally adapted to the numeration, the multiplication, or the division, of material substances, either in his own person, or in external nature. The proportions of the human body, and of its members, are in other than decimal numbers. The first unit of measures, for the use of the hand, is the *cubit*, or extent from the tip of the elbow to the end of the middle finger; the motives for choosing which, are, that it presents more definite terminations at both ends than any of the other superior limbs, and gives a measure easily handled and carried about the person. By doubling this measure is given the ell, or arm, including the hand, and half the width of the body, to the middle of the breast; and, by doubling that, the fathom, or extent from the extremity of one middle finger to that of the other, with expanded arms, an exact equivalent to the stature of man, or extension from the crown of the head to the sole of the foot. For subdivisions and smaller measures, the span is found equal to half the cubit, the palm to one third of the span, and the finger to one fourth of the palm. The cubit is thus, for the mensuration of matter, naturally divided into 24 equal parts, with subdivisions of which 2, 3, and 4, are the factors; while, for the mensuration of distance, the foot will be found at once equal to one fifth of the pace, and one sixth of the fathom.

Nor are the diversities of nature, in the organization of external matter, better suited to the exclusive use of decimal arithmetic. In the three modes of its extension, to which the same linear measure may be applied, length, breadth, and thickness, the proportions of surface and solidity are not the same with those of length: that which is decimal to the line, is centesimal to the surface, and millesimal to the cube. Geometrical progression forms the rule of numbers for the surface and the solid, and their adaptation to decimal numbers is among the profoundest mysteries of mathematical science, a mystery which had been impenetrable to Pythagoras, Archimedes, and Ptolemy; which remained unrevealed even to Copernicus, Galileo, and Kepler, and the discovery and exposition of which was reserved to immortalize the name of Napier. To the mensuration of the surface and the solid, the number ten is of little more use than any other. The numbers of each of the two or three modes of extension must be multiplied together to yield the surface or the solid contents: and, unless the object to be measured is a perfect square or cube of equal dimensions at all its sides, decimal arithmetic is utterly incompetent to the purpose of their admeasurement.

Linear measure, to whatever modification of matter applied, extends in a straight line; but the modifications of matter, as produced by nature, are in forms innumerable, of which the defining outward line is almost invariably a curve. If decimal arithmetic is incompetent even to give the dimensions of those artificial forms, the square and the cube, still more incompetent is it to give the circumference, the area, and the contents, of the circle and the sphere.

There are three several modes by which the quantities of material substances may be estimated and compared; by number, by the space which they occupy, and by their apparent specific gravity. We have seen the origin and character of mensuration by space and number, and that, in the order of human existence, one is the result of a necessity incidental to individual man preceding the social union, and the other immediately springing from that union. The union of the sexes constitutes natural society: their permanent cohabitation is the foundation of domestic society, and leads to that of government, arising from the relations between the parents and the offspring which their union produces. The relations between husband and wife import domestic society, consent, and the sacred obligation of promises. Those between parent and child, import subordination and government; on the one side authority, on the other obedience. In the first years of infancy, the authority of the parent is absolute; and has, therefore, in the laws of nature, been tempered by parental affection. As the child advances to mature age, the relations of power and subjection gradually subside, and, finally, are dissolved in that honor and reverence of the child for the parent, which can terminate only with life. When the child goes forth into the world to make a settlement for himself, and found a new family, civil society commences; government is instituted—the tillage of the ground, the discovery and use of metals, exchanges, traffic by barter, a *common* standard of measures, and mensuration by *weight*, or apparent specific gravity, all arise from the multiplying relations between man and man, now superadded to those between man and things.

The difference between the specific gravities of different substances is so great, that it could not, for any length of time, escape observation; but nature has not furnished man, within himself, with any standard for this mode of estimating equivalents. Specific gravity, as an object of mensuration, is in its nature *proportional*. It is not like measures of length and capacity, a comparison between different definite portions of space, but a comparison between different properties of matter. It is not the simple relation between the extension of one substance, and the extension of another; but the complicated relation of extension and gravitation in one substance to the extension and gravitation of another. This distinction is of great and insuperable influence upon the principle of *uniformity*, as applicable to a system of weights and measures. *Extension* and *gravitation* neither have, nor admit of, one common standard. *Diversity* is the law of their nature, and the only *uniformity* which human ingenuity can establish between them is, an uniformity of proportion, and not an uniformity of identity.

The necessity for the use of *weights* is not in the organization of individual man. It is not essential even to the condition or the comforts of domestic society. It presupposes the discovery of the properties of the balance; and originates in the exchanges of traffic, after the institution of civil society. It results from the experience that the comparison of the articles of exchange, which serve for the subsistence or the enjoyment of life, by their relative extension, is not sufficient as a criterion of their value. The first use of the balance, and of weights, implies two substances, each of which is the test and the standard of the other. It is natural that these substances should be the articles the most essential to subsistence. They will be borrowed from the harvest and the vintage: they will be corn and wine. The discovery of the metals, and their extraction from the bowels of the earth, must, in the annals of human nature, be subsequent, but proximate, to the first use of weights; and, when discovered, the only mode of ascertaining their definite quantities will be soon perceived to be their weight. That they should, themselves, immediately become the common standards of exchanges, or otherwise of value and of weights, is perfectly in the order of nature; but their proportions to one another, or to the other objects by which they are to be estimated, will not be the same as standards of weight, and as standards of value. Gold, silver, copper, and iron, when balanced each by the other in weight, will present masses very different from each other in value. They give rise to another complication, and another diversity, of weights and measures, equally inaccessible to the uniformity of identity, and to the computations of decimal arithmetic.

Of the metals, that which, by the adaptation of its properties to the various uses of society, and to the purposes of traffic, by the quantities in which nature has disclosed it to the possession of man, intermediate between her profuse bounties of the coarser, and her parsimonious dispensation of the finer, metals, holds a middle station between them, wins its way as the common, and at last as the only, standard of value. It becomes the universal medium of exchanges. Its quantities, ascertained by weight, become themselves the standards of weights. Civil government is called in as the guardian and voucher of its purity. The civil authority stamps its image, to authenticate its weight and alloy: and silver becomes at once a weight, money, and coin.

With civil society too originates the necessity for common and uniform standards of measures. Of the different measures of extension necessary for individual man, and for domestic society, although the want will be common to all, and frequently recurring, yet, the standards will not be uniform, either with reference to time or to persons.

The standard of linear measure for each individual being in himself, those of no two individuals will be the same. At different times, the same individual will use different measures, according to the several purposes for which they will be wanted. In domestic society, the measures adaptable to the persons of the husband, of the wife, and of the children, are not the same; nor will the idea of reducing them all to one common standard press itself upon their wants, until the multiplication of families gives rise to the intercourse, exchanges, and government, of civil society. Common standards will then be assumed from the person of some distinguished individual; but accidental circumstances, rather than any law of nature, will determine whether identity or proportion will be the character of their uniformity. If, pursuing the first and original dictate of nature, the cubit should be assumed as the standard of linear measure for the use of the hand, and the pace for the measure of motion, or linear measure upon earth, there will be two units of long measure; one for the measure of matter, and another for the measure of motion. Nor will they be reducible to one; because neither the cubit nor the pace is an aliquot part or a multiple of the other. But, should the discovery have been made, that the *foot* is at once an aliquot part of the pace, for the mensuration of motion, and of the ell and fathom, for the mensuration of matter, the foot will be made the common standard measure for both: and, thenceforth, there will be only one standard unit of long measure, and its uniformity will be that of identity.

Thus, in tracing the theoretic history of weights and measures to their original elements in the nature and the necessities of man, we have found linear measure with individual existence, superficial, capacious, itinerary measure, and decimal arithmetic, with domestic society; weights and common standards, with civil society; money, coins, and all the elements of uniform metrology, with civil government and law; arising in successive and parallel progression together.

When weights and measures present themselves to the contemplation of the legislator, and call for the interposition of law, the first and most prominent idea which occurs to him is that of *uniformity:* his first object is to embody them into a system, and his first wish, to reduce them to one universal common standard. His purposes are uniformity, permanency, universality; one standard to be the same for all persons and all purposes, and to continue the same forever. These purposes, however, require powers which no legislator has hitherto been found to possess. The power of the legislator is limited by the extent of his territories, and the numbers of his people. His principle of universality, therefore, cannot be made, by the mere agency of his power, to extend beyond the inhabitants of his own possessions. The power of the legislator is limited over time. He is liable to change his own purposes. He is not infallible: he is liable to mistake the means of effecting his own objects. He is not immortal: his successor accedes to his power, with different views, different opinions, and perhaps different principles. The legislator has no power over the properties of matter. He cannot give a new constitution to nature. He cannot repeal her law of universal mutability. He cannot square the circle. He cannot reduce extension and gravity to one common measure. He cannot divide or multiply the parts of the surface, the cube, or the sphere, by the uniform and exclusive number ten. The power of the legislator is limited over the will and actions of his subjects. His conflict with them is desperate, when he counteracts their settled habits, their established usages; their domestic and individual economy, their ignorance, their prejudices, and their wants: all which is unavoidable in the attempt radically to change, or to originate, a totally new system of weights and measures.

In the origin of the different measures and weights, at different stages of man's individual and social existence; in the different modes by which nature has bounded the extension of matter; in the incommensurable properties of the straight and the curve line; in the different properties of matter, number, extension, and gravity, of which measures and weights are the tests, nature has planted sources of diversity, which the legislator would in vain overlook, which he would in vain attempt to control. To these sources of diversity in the nature of things, must be added all those arising from the nature and history of man. In the first use of weights and measures, neither universality nor permanency are essential to the uniformity of the standards. Every individual may have standards of his own, and may change them as convenience or humor may dictate. Even in civil society, it is not *necessary*, to the purposes of traffic, that the standards of the buyer and seller should be the same. It suffices, if the proportions between the standards of both parties are mutually understood. In the progress of society, the use of weights and measures having preceded legislation, if the families, descended from one, should, as they naturally may, have the same standards, other families will have others. Until regulated by law, their diversities will be numberless, their changes continual.

These diversities are still further multiplied by the abuses incident to the poverty, imperfections, and deceptions, of human language. So arbitrary and so irrational is the dominion of usage over the speech of man, that, instead of appropriating a specific name to every distinct thing, he is impelled, by an irresistible propensity, sometimes to gives different names to the same thing, but far more frequently to give the same name to different things. Weights and measures are, in their nature, relative. When man first borrows from his own person a standard measure of length, his first error is to give to the measure the name of the limb from which it is assumed. He calls the *measure* a cubit, a span, a hand, a finger, or a foot, improperly applying to it the name of those respective parts of his body. When he has discovered the properties of the balance, he either confounds with it the name of the weight, which he puts in it to balance the article which he would measure, or he gives to the definite mass, which he assumes for his standard, the indefinite and general name *of the weight.* Such was the original meaning of the weight which we call a *pound*. But, as different families assume different masses of gravity for their unit of weight, the pound of one bears the same name, and is a very different thing from the pound of another. When nations fall into the use of different weights or measures for the estimation of different objects, they commit the still grosser mistake of calling several different weights or measures by the same name. And, when governments degrade themselves by debasing their coins, as unfortunately all governments have done, they add the crime of fraud to that of injustice, by retaining the name of things which they have destroyed or changed. Even things which nature has discriminated so clearly, that they cannot be mistaken, the antipathy of mankind to new words will misrepresent and confound. It suffers not even numbers to retain their essentially definite character. It calls sixteen a dozen. It makes a hundred and twelve a hundred, and twenty-eight, twenty-five. Of all the tangles of confusion to be unravelled by the regulation of weights and measures, these abuses of language in their nomenclature are perhaps the most inextricable. So that when law comes to establish its principles of permanency, uniformity, and universality, it has to contend not only with the diversities arising from the nature of things and of man; but with those infinitely more numerous which proceed from existing usages, and delusive language; with the partial standards, and misapplied names, which have crept in with the lapse of time, beginning with individuals or families, and spreading more or less extensively to villages and communities.

In this conflict between the dominion of usage and of law, the last and greatest dangers to the principle of uniformity proceed from the laws themselves. The legislator having no distinct idea of the uniformity of which the subject is susceptible, not considering how far it should be extended, or where it finds its boundary in the nature of things and of man, enacts laws inadequate to their purpose, inconsistent with one another; sometimes stubbornly resisting, at others weakly yielding to inveterate usages or abuses; and finishes by increasing the diversities which it was his intention to abolish, and by loading his statute book only with the impotence of authority, and the uniformity of confusion.

This inquiry into the theory of weights and measures, as resulting from the natural history of man, was deemed necessary as preliminary to that statement of the proceedings of foreign countries for establishing uniformity in weights and measures, called for by the resolution of the Senate.

It presents to view certain principles believed to be essential to the subject, upon which the historical statement required will shed continual illustration, and which it will be advisable to bear in mind, when the propositions supposed to be proper for the adoption of the United States are to be considered.

In this review, civil society has been considered as originating in a single family. It can never originate in any other manner. But government, and national communities, may originate, either by the multiplication of families from one, or in compact, by the voluntary association of many families, or in force, by conquest. In the nations formed by the reunion of many families, each family will have its standard measures and weights already settled, and common standards for the whole can be established only by the means of *law*. It is a consideration from which many important consequences result, that the proper province of law, in relation to weights and measures, is, not to create, but to regulate. It finds them already existing, with diversities innumerable, arising not only from all the causes which have been enumerated, but from all the frauds to which these diversities give continual occasion and temptation.

There are two nations of antiquity from whom almost all the civil, political, and religious institutions of modern Europe, and of her descendants in this hemisphere, are derived—the Hebrews, and the Greeks. They both, at certain periods, not very distant from each other, issued from Egypt; and both nearly at the time of the first invention of alphabetical writing. The earliest existing records of history are of them, and in their respective languages. They exhibit examples of national communities and governments originating in two of the different modes noticed in the preceding remarks. The Hebrews sprung from a single family, of which Abraham and Sarah were the first founders. The Greeks were a confederated nation, formed by the voluntary association of many families. To their historical records, therefore, we must appeal for the actual origin of our own existing weights and measures; and, beginning with the most ancient of them, the Hebrews, it is presumed, that the scriptures may be cited in the character of historical documents. We there find, that all the human inhabitants of this globe sprung from one created pair; that the necessity of raiment adapted to the organization of their bodies, and of the tillage of the ground for their subsistence, arose by their fall from innocence; that their eldest son was a tiller of the ground, and built a city, and their second son a keeper of sheep; that, at no distant period from the creation, instruments of brass and iron were invented. Of the origin of weights and measures, no direct mention is made; but the Hebrew historian, Josephus, asserts, that they were invented by Cain, the tiller of the ground, and the first builder of a city. As the duration of human life was tenfold longer before the flood than in later ages, the multiplication of the species was proportionally rapid; and the inventions and discoveries of many ages were included within the life of every individual. In the early stages of man's existence upon earth, direct revelations from the Creator were also frequent, and imparted knowledge unattainable but in a series of centuries to the merely natural energies of the human mind. The division of numbers by decimal arithmetic, and the use of the *cubit* as a standard measure of length, are distinctly proved to have been established before the general deluge. The division of time into days, months, and years, was settled. The ages of the patriarchs are noted in units, tens, and hundreds of years; and

Noah, we are told, built, by divine instruction, his ark three hundred *cubits* long, fifty cubits broad, and thirty cubits in height.

After the general deluge, the dispersion of the human species, and the confusion of languages which ensued, must have destroyed whatever uniformity of weights and measures might have existed, while the whole earth was of one language and of one speech. After noticing this great and miraculous event, the historical part of the Bible is chiefly confined to the family of Abraham, originally a Chaldean, said to have been very rich in cattle, in silver, and in gold. In his time, we find mention made of *measures* of meal. Abimelech gives him a thousand pieces of silver. He, himself gives to Hagar a *bottle* of water, and buys of Ephron, the Hittite, the field of Machpelah, for which he pays him, by *weight*, four hundred shekels of silver, current money with the merchant. At this period, therefore, we find established measures of length, of land, and of capacity, liquid and dry; weights, coined money, and decimal arithmetic. The elements for a system of metrology are complete; but the only uniformity observable in them is, the identity of weights and coin, and the decimal numbers.

In the *law* given from Sinai—the law, not of a human legislator, but of God—there are two precepts respecting weights and measures. The first, [Leviticus xix. 35, 36] "Ye shall do no unrighte-"ousness in judgment, in mete-yard (measure of length) in weight, "or in measure (of capacity). Just balances, just weights, a just "ephah, and a just hin, shall ye have." The second, [Deuteronomy xxv. 13, 14, 15] "Thou shalt not have in thy bag *divers* weights, "a great and a small. Thou shalt not have in thine house divers "measures, a great and a small. But thou shalt have a perfect and "just weight, a perfect and just measure shalt thou have." The weights and measures are prescribed as already existing and known, and were all probably the same as those of the Egyptians. The first of these injunctions is addressed in the plural to the whole nation, and the second in the singular to every individual. The first has reference to the standards, which were to be kept in the ark of the covenant, or the sanctuary; and the second to the copies of them, kept by every family for their own use. The first, therefore, only commands that the standards should be *just:* and that, in all transactions, for which weights and measures might be used, the principle of righteousness should be observed. The second requires, that the copies of the standards used by individuals, should be uniform, not divers; and not only just, but perfect, with reference to the standards.

The long measures were, the *cubit*, with its subdivisions of two *spans*, six palms or hand-breadths, and twenty-four digits or fingers. It had no division in decimal parts, and was not employed for itinerary measure: that was reckoned by paces, Sabbath day's journeys, and day's journeys. The measures of capacity were, the ephah for the dry, and the *hin* for liquid measure; the primitive standard from nature of which was an egg-shell; six of these, constituted the *log*, a measure little less than our *pint*. The largest measure of capacity, the *homer*, was common both to liquid and dry substances; and its contents nearly corresponded with our wine hogshead, and with the Winchester quarter. The intermediate measures were different, and differently subdivided. They combined the decimal and duodecimal divisions: the latter of which may, perhaps, have arisen from the accidental number of the tribes of Israel. Thus, in liquids, the bath was a tenth part of the homer, the hin a sixth part of the bath, and the log a twelfth part of the hin; while, for dry measure, the ephah was a tenth part of the homer, the seah a third, and the omer a tenth part of the ephah, and the cab a sixth part of the seah. The weights and coins were, the shekel, of twenty gerahs; the maneh, which for weight was of sixty and in money of fifty shekels; and the kinchar, or talent, of three thousand shekels in both. The ephah had also been formed by the process of cubing an Egyptian measure of length, called the *ardob*. The original weight of the shekel was the same as one-half of our avoirdupois ounce; the most ancient of weights traceable in human history.

And thus the earliest and most venerable of historical records extant, in perfect coincidence with speculative theory, prove, that the natural standards of weights and measures are not the same; that even the natural standards of cloth and of long measure are two, both derived from the stature and proportions of man, but one from his hand and arm, the other from his leg and foot; that the natural standards of measures of capacity and of weights are different from those of linear measure, and different from each other, the essential character of the weight being compact solidity, and that of the vessel bounded vacuity; that the natural standards of weights are two, one of which is the same with metallic money; and that decimal arithmetic, as founded in nature, is peculiarly applicable to the standard *units* of weights and measures, but not to their subdivisions or fractional parts, nor to the objects of admeasurement and weight.

With all these diversities, the only commands of the law for observing uniformity were, that the weights and the measures should be just, perfect, and not divers, a *great* and a *small*. But this last prohibition was merely an ordinance against fraud. It was a precept to the individual, and not to the nation. It forbade the iniquitous practice of using a large weight or measure for buying, and a small one for selling the same article; and, to remove the opportunity for temptation, it enjoined upon the individual not to have divers weights and measures, great and small, of the same denomination, in his bag when at market, or in his house when at home. But it was never understood to forbid that there should be measures of different dimensions bearing the same name: and it appears, from the sacred history, that there actually were three different measures called a cubit, of about the relative proportion of 17, 21, and 35, of our inches, to each other. They were distinguished by the several denominations of the cubit of a man, the cubit of the king, and the cubit of the sanctuary.

ON WEIGHTS AND MEASURES.

In the vision of the prophet Ezekiel, during the Babylonian captivity—that vision which, under the resurrection of dead bones, shadowed forth the restoration and union of the houses of Ephraim and of Judah, with the reproaches of former violence and spoil, injustice and exactions, are mingled the exhortations of future righteousness, particularly with reference to weights and measures: and there is a special command that the measures of capacity, liquid and dry, should be the same.

"Thus saith the Lord God; let it suffice you, O princes of Israel, remove violence and spoil, and execute judgment and justice, take away your exactions from my people, saith the Lord God. Ye shall have just balances, and a just ephah, and a just bath. The ephah and the bath shall be one measure, that the bath may contain the tenth part of an homer, and the ephah the tenth part of an homer; the measure thereof shall be after the homer. And the shekel shall be twenty gerahs: twenty shekels, five and twenty shekels, fifteen shekels, shall be your maneh. This is the oblation that ye shall offer; the sixth part of an ephah of an homer of wheat; and ye shall give the sixth part of an ephah of an homer of barley. Concerning the ordinance of oil, ye shall offer the tenth part of a bath out of the cor, which is an homer of ten baths; for ten baths are an homer."

Here we see combined the uniformity of identity, and the uniformity of proportion. The homer was a dry, and the cor a liquid, measure of capacity: they were of the same contents: the ephah and the bath were their corresponding tenth parts, also of the same capacity. But the oblation of wheat and barley was to be a *sixth* part of the ephah, and the oblation of oil a tenth part of the bath. The oblations were uniform, but the measures were proportional; and that proportion was compounded of the different weight and value of the respective articles.

In the same vision of Ezekiel, the directions are given for the building of the new temple after the restoration of the captivity; and all the dimensions of the temple are prescribed by a measuring reed of six cubits long by the cubit and an hand-breadth. "And these (says he) are the measures of the altar after the cubits: *the cubit is a cubit and an hand-breadth.*" [Ch. xliii, 13.]

The book of Job is a story of a man, supposed not to have been descended from Abraham, and certainly not belonging to any of the tribes of Israel. It has reference to other manners, other customs, opinions, and laws, than those of the Hebrews. But it bears evidence of the primitive custom of paying silver by weight, while gold and jewels were valued by tale; and of that system of proportional uniformity, which combines gravity and extension for the mensuration of fluids. Speaking of wisdom, it says, [ch. xxviii, v. 15, 17] "It cannot be gotten for gold, neither *shall silver be weighed* for the price thereof. The gold and the crystal cannot equal it; and the exchange of it shall not be for jewels of fine gold." And, afterwards, in the same chapter, that " God maketh the weight for the " winds, and *and weigheth the waters by measure.*"

The cubit was also a primitive measure of length among the Greeks; but, at the institution of the Olympic games, by Hercules, his foot is said to have been substituted as the unit of measure for the foot-race. Six hundred of these feet constituted the stadium, or length of the course or stand, which thenceforth became the standard itinerary measure of the nation. It was afterwards by the Romans combined with the pace, a thousand of which constituted the mile. The foot and the mile, or thousand paces, are our standard measures of length at this day.

The foot has over the cubit the advantage of being a common aliquot part both of the pace and the fathom. It is also definite at both extremities, and affords the natural means of reducing the two standard measures of length to one. Its adoption was therefore a great and important advance towards uniformity: and this may account for the universal abandonment, by all the modern nations of Europe, of that primitive antediluvian standard measure, the *cubit*.

Of the Greek weights and measures of capacity, the origin is not distinctly known; but that whatever uniformity ever existed in the system was an uniformity of proportion, and not of identity, is certain. They had weights corresponding to our avoirdupois and troy pounds, and measures answering to our wine and ale gallons; not indeed in the same proportions; but in the proportions to each other of the weight of wine and oil.

It has been observed that the process of weighing implies two substances, each of which is the standard and test of the other; that, in the order of human existence, the use of weights precedes the weighing of metals, but that when the metals and their uses to the purposes of life are discovered, their value can at first be estimated only by weight, whereby they soon become standards both of weight and of value for all other things. This theory is confirmed by the history of the Greek, no less than by that of the Hebrew, weights and measures. The term talent, in its primitive meaning in the Greek language, signified a *balance;* and it was at once the largest weight and the highest denomination of money among the Greeks. Its subdivisions, the *mina,* and the *drachma,* were at once weights and money; and the *drachma* was the unit of all the silver coins. But the money which was a weight, though substituted for many purposes, instead of the more ancient weight by which it had itself been tried, never excluded it from use. It had not the fortune of the *foot,* to banish from the use of mankind its predecessor. They had the weight for money, and the weight for measure.

As there are thus in nature two standards of weight, there are also two of measures of capacity. From the names of the Greek measures of capacity, they were originally assumed from cockle and other shells of fish. But as these give no scales of proportion for subdivisions, when reduced to a system, their capacity was determined by the two modifications of matter, extension and weight.

Like the Hebrews, they had measures for liquid and dry substances which were the same; but with different multiples and subdivisions. Their measures of wine and oil were determined by the *weight* of their contents; their measures of water and of grain, by vessels of capacity cubed from measures of length.

The weights and measures of the Romans were all derived from those of the Greeks. The identity of one of their standard units of weight, with money and coin, was the same. *Aes*, brass, was their original money: and as its payment was by weight, the term pound, *libra*, was the balance; and *money* was the weight of brass in the balance. The general term soon came to be applied to a definite weight: and when afterwards silver came to be coined, the *sestertius*, which signified two and a half, and the *denarius*, or piece of *ten*, meant the pieces of silver of value equal respectively to two and a half and to ten of the original brass weights of the balance. The sestertius was the unit of money, and the denarius of silver coins.

The Romans had also two pound weights; which were termed the metrical and the scale pound. "The scale pound," says Galen, "determines the *weight* of bodies; the metrical pound, the contents "or quantity of *space* which they fill."

Their measures of capacity for wet or dry substances were in like manner, in part, the same, but with different multiples and subdivisions. Like them they were formed of the two different processes of cubing the foot, and of testing wine and oil by weight. The *amphora*, or largest measure of liquids, weighed eighty pounds of water, and was formed by cubing, or, as they called it, squaring their foot measure: it was for that reason called a *quadrantal*. But their *congius*, or unit of liquid measure, was any vessel containing ten metrical *pounds weight* of wine. The Silian law, enacted nearly three centuries before the Christian era, expressly declares that the quadrantal contains eighty *pounds* of wine, the *congius* ten pounds; that the *sextarius* contains the sixth *part* of a congius, and is a measure both for liquid and dry substances; that forty-eight sextarii make a quadrantal of wine, and sixteen *libræ* a modius. The money pound, or *pondo*, and the metrical pound, or *libra*, were in the proportion to each other of 84 to 100, nearly the same as that between our troy and avoirdupois weights. [Arbuthnot on Coins, Weights, and Measures, p. 23.] There is a standard congius of the age of Vespasian still extant at Rome; and the inscription upon it marks, that it contains ten pounds of wine.

Among the nations of modern Europe there are two, who, by their genius, their learning, their industry, and their ardent and successful cultivation of the arts and sciences, are scarcely less distinguished than the Hebrews from whom they have received most of their religious, or the Greeks from whom they have derived many of their civil and political institutions. From these two nations the inhabitants of these United States are chiefly descended; and from one of them we have all our existing weights and measures. Both of them, for a series of ages, have been engaged in the pursuit of an uniform system of weights and measures. To this the wishes of their philanthropists, the hopes of their patriots, the researches of their philosophers, and the energy of their legislators, have been aiming with efforts so stupendous and with perseverance so untiring, that, to any person who shall examine them, it may well be a subject of astonishment to find that they are both yet entangled in the pursuit at this hour, and that it may be doubted whether all their latest and greatest exertions have not hitherto tended to increase diversity instead of producing uniformity.

It was observed, at the introduction of these remarks, that one of the primary elements of uniformity, as applied to a system of weights and measures, has reference to the *persons* by whom they are used; and it has since been noticed, that the power of the legislator is restricted to the inhabitants of his own dominions. Now, the perfection of uniformity with respect to the persons to whose use a system of metrology is adapted, consists in its embracing, at least in its aptitude, the whole human race. In the abstract, that system which would be most useful for one nation, would be the best for all. But this uniformity cannot be obtained by legislation. It must be imposed by conquest, or adopted by consent. When therefore two populous and commercial nations are at the same time forming and maturing a system of weights and measures on the principle of uniformity, unless the system proves to be the same, the result as respects all their relations with each other must be, not uniformity, but new and increased diversity. This consideration is of momentous importance to the people of this Union. Since the establishment of our national independence, we have partaken of that ardent spirit of reform, and that impatient longing for uniformity, which have so signally animated the two nations from whom we descended. The Congress of the United States have been as earnestly employed in the search of an uniform system of weights and measures as the British Parliament. Have either of them considered, how that very principle of uniformity would be affected by any, the slightest change, sanctioned by either, in the existing system, now common to both? If uniformity be their object, is it not necessary to contemplate it in all its aspects? And while squaring the circle to draw a straight line from a curve, and fixing mutability to find a standard pendulum, is it not worth their while to inquire, whether an imperceptible improvement in the uniformity of things would not be dearly purchased by the loss of millions in the uniformity of persons?

It is presumed that the intentions of the Senate, in requiring a statement of the proceedings in foreign countries for establishing uniformity in weights and measures will be fulfilled by confining this part of the inquiry to the proceedings of the two nations above mentioned. It appears that a reformation of the weights and measures of Spain is among the objects now under the consideration of the Cortes of that kingdom ; and, as weights and measures are the necessary and universal instruments of commerce, no change can be effected in the system of any one nation without sensibly affecting, though in very different degrees, all those with whom they entertain any relations of trade. But the results of this inquiry, newly instituted in Spain, have not yet been made known. France and Great Britain are the only nations of modern Europe who have taken much interest in the organization of a new system, or attempted a reform for the avowed purpose of uniformity. The proceedings in those two countries have been numerous, elaborate, persevering, and, in France especially, comprehensive, profound, and systematic. In both, the phenomenon is still exhibited, that, after many centuries of study, of invention of laws, and of penalties, almost every village in the country is in the habitual use of different weights and measures; which diversity is infinitely multiplied, by the fact, that, in each country, although the quantities of the weights and measures are thus different, their denominations are few in number, and the same names, as foot, pound, ounce, bushel, pint, &c. are applied in different places, and often in the same place, to quantities altogether diverse.

During the conquering period of the French Revolution, the new system of French weights and measures was introduced into those countries which were united to the empire. Since the severance of those countries from France, it has been discarded, excepting in the kingdom of the Netherlands, where, by two ordinances of the king, it has been confirmed with certain exceptions and modifications, particularly with regard to the coins.

In England, from the earliest records of parliamentary history, the statute books are filled with ineffectual attempts of the legislature to establish uniformity. Of the origin of their weights and measures, the historical traces are faint and indistinct: but they have had, from time immemorial, the pound, ounce, foot, inch, and mile, derived from the Romans, and through them from the Greeks, and the *yard*, or *girth*, a measure of Saxon origin, derived, like those of the Hebrews and the Greeks, from the human body, but, as a natural standard, different from theirs, being taken not from the length or members, but from the circumference of the body. The yard of the Saxons evidently belongs to a primitive system of measures different from that of the Greeks, of which the foot, and from that of the Hebrews, Egyptians, and Antediluvians, of which the cubit was the standard. It affords, therefore, another demonstration, how invariably nature first points to the human body, and its proportions, for the original standards of linear measure. But the *yard* being for all purposes of use a measure corresponding with the *ulna*, or ell, of the Roman system, became, when superadded to it, a source of diversity, and an obstacle to uniformity in the system. The yard, therefore, very soon after the Roman conquest, is said to have lost its original character of girth; to have been adjusted as a standard by the arm of king Henry the First: and to have been found or made a multiple of the foot, thereby adapting it to the remainder of the system: and this may perhaps be the cause of the difference of the present English foot from that of the Romans, by whom, as a measure, it was introduced. The ell measure has, however, in England, retained its place as a standard for measuring cloth : but, in the ancient statutes, which for centuries after the conquest were enacted in the degenerate Latin of the age, the term *ulna,* or ell, is always used to designate the yard. Historical traditions allege that, a full century before the Conquest, a law of Edgar prescribed that there should be the same weights, and the same measures, throughout the realm ; but that it was never observed. The system which had been introduced by the Romans, however uniform in its origin, must have undergone various changes in the different governments of the Saxon Heptarchy. When those kingdoms were united in one, it was natural that laws of uniformity should be prescribed by the prince ; and as natural that usages of diversity should be persisted in by the people. Canute the Dane, William the Conqueror, and Richard the First, princes among those of most extensive and commanding authority, are said to have made laws of the like import, and the same inefficacy. The Norman Conquest made no changes in any of the established weights and measures. The very words of a law of William the Conqueror are cited by modern writers on the English weights and measures ; their import is : " We ordain and command that the weights and measures, " throughout the realm, be as our worthy predecessors have estab-" lished." [Wilkins, Legg, Saxon, Folkes, cited by Clark, p. 150.]

One of the principal objects of the Great Charter was the establishment of uniformity of weights and measures ; but it was a uniformity of *existing* weights and measures ; and a uniformity not of identity, but of proportion. The words of the 25th chapter of the Great Charter of the year 1225 (9 Henry III.) are, in the English translation of the statutes, " one measure of wine shall be through our realm, " and one measure of ale, and one measure of corn, that is to say, " the *quarter* of London : and one breadth of dyed cloth, that " is to say, two yards [ulne] within the lists : and it shall be of " weights as it is of measures." The London quarter, therefore, and the *yard,* or *ulna,* were existing, known, established measures ; and the one measure of corn was the London quarter. The one measure of ale was a gallon, of the same contents for liquid measure as the half peck was for dry. But the one measure of wine was a gallon, not of the same cubical contents as the half-peck and ale gallon, but which, when filled with wine, was of the same weight as the half-peck, or corn gallon, when filled with wheat. And the expressions, " it shall be of weights as it is of measures," mean that there shall be the same proportion between the money weight, and the merchant's weight, as between the wine measure and the corn measure.

The Great Charter, which now appears as the first legislative act in the English statutes at large, is not the Magna Charta extorted by the barons from John, at Runnimead, but a repetition of it by Henry the Third in the year 1225, as confirmed by his son, Edward the First, in the year 1300. It is properly an act of this last date, though inserted in the book as of 9 Henry III., or 1225.

In several of the subsequent confirmations of this charter, which, for successive ages, attest at once how apt it was to be forgotten by power, and how present it always was to the memory of the people, the real meaning of this 25th chapter appears to have been misunderstood. It has been supposed to have prescribed the uniformity of identity, and not the uniformity of proportion; that, by enjoining one measure of wine, and one measure of ale, and one measure of corn, its intention was, that all these measures should be the same; that there should be only one unit measure of capacity for liquid and dry substances, and one unit of weights.

But this neither was, nor could be, the meaning of the statute. Had it been the intention of the legislator, he would have said, there shall be one and the same measure for wine, corn, and ale; and the reference to the London quarter could not have been made, for neither wine nor ale were ever measured by the quarter; and, instead of saying "it shall be of weights as it is of measures," it would have said, there shall be but one set of weights for whatever is to be weighed.

The object of the whole statute was, not to innovate, but to fix existing rights and usages, and to guard against fraud and oppression. It says that the measure of corn shall be the London quarter; that cloth shall be two yards within the lists. But it neither defines the contents of the quarter, nor the length of the yard: it refers to both as fixed and settled quantities. To have prescribed that there should be but one unit of weights and one measure of wine, ale, and corn, would have been a great and violent innovation upon all the existing habits and usages of the people. The chapter is not intended for a *general* regulation of weights and measures. It refers specifically and exclusively to the measure of three articles, wine, ale, corn; and to the width of cloths. Its intention was to provide that the measure of corn, of ale, and of wine, should *not* be the same; that is, that the wine measure should not be used for ale and corn, nor the ale measure for wine.

That such was and must have been the meaning of the statute, is further proved by the statute of 1266, (51 Henry III.) and by the treatise upon weights and measures, published in the statute books as of the 31 Edward I., or 1304; the first, an act of the same Henry the Third whose Great Charter is that inserted among the laws, and the second an act of the same Edward the First whose confirmation of the Great Charter is the existing statute.

The act of 51 Henry III., (1266) is called the assize of bread and of ale. It purports to be an exemplification, given at the request of the bakers of the town of Coventry, of certain ordinances, of the assize of bread, and ale, and of the making of money and measures, made in the times of the king's progenitors, sometime kings of England. It presents an established scale, then of ancient standing, between the prices of wheat and of bread, providing that when the *quarter* of wheat is sold at twelve pence, the farthing loaf of the best white bread shall weigh six pounds sixteen shillings. It then graduates the weight of bread according to the price of wheat, and for every six pence added to the quarter of wheat, reduces, though not

in exact proportions, the weight of the farthing loaf, till, when the wheat is at twenty shillings a quarter, it directs the weight of the loaf to be six shillings and three pence. It regulates, in like manner, the price of the gallon of ale, by the price of wheat, barley, and oats; and, finally, declares, that, " by the consent of *the whole realm* of Eng-
" land, the *measure* of the king was made; that is to say: that an
" English *penny*, called a sterling round, and without any clipping,
" shall weigh thirty-two wheat corns in the midst of the ear, and
" twenty-pence do make an ounce, and twelve ounces one pound, and
" eight pound do make a gallon of *wine*, and eight gallons of *wine*
" do make a London bushel, which is the eighth part of a quarter."

Henry the Third was the eighth king of the Norman race: and this statute was passed exactly two hundred years after the Conquest. It is merely an exemplification, word for word, embracing several ordinances of his progenitors, kings of England; and it unfolds a system of uniformity for weights, coins, and measures of capacity, very ingeniously imagined, and skilfully combined.

It shows, first, that the money weight was identical with the silver coins: and it establishes an uniformity of proportion between the money weight and the merchant's weight, exactly corresponding to that between the measure of wine and the measure of grain.

It makes wheat and silver money, the two weights of the balance, the natural tests and standards of each other; that is, it makes wheat the standard for the weight of silver money, and silver money the standard for the weight of wheat.

It combines an uniformity of proportion between the weight and the measure of wheat and of wine; so that the measure of wheat should at the same time be a certain weight of wheat and the measure of wine at the same time a certain weight of wine, so that the article whether bought and sold by weight or by measure, the result was the same. To this, with regard to wheat, it gave the further advantage of an abridged process for buying or selling it by the number of its kernels. Under this system, wheat was bought and sold by a combination of every property of its nature, with reference to quantity; that is, by number, weight, and measure. The statute also fixed its proportional weight and value with reference to the weight and value of the silver coin for which it was to be exchanged in trade. If, as the most eminent of the modern economists maintain, the value of every thing in trade is regulated by the proportional value of money and of wheat, then the system of weights and measures, contained in this statute, is not only accounted for as originating in the nature of things, but it may be doubted whether any other system be reconcileable to nature. It was with reference to this system, that, in the introduction to this report, it was observed, that our own weights and measures were originally founded upon an uniformity of proportion, and not upon an uniformity of identity. In the system which allows only one unit of weights and one unit of measures of capacity, all the advantages of the uniformity of proportion are lost. The litre of the French system is a weight for nothing but distilled water, at a given temperature.

But with this statute of 1266, and with the admirable system of proportional uniformity in weights and measures, of which it gives the elements, it has fared still worse than with the twenty-fifth chapter of Magna Charta. The most valuable and important feature of uniformity in the system, the identity of the nummulary weight and of the standard silver coin, that feature which is believed to be of more influence upon the happiness and upon the morals of nations, than any other principle of uniformity of which weights and measures are susceptible, was first defaced by Edward the First himself. It was utterly annihilated by his successors. The consequence of which has been, that the object and scope of the statute of 1266 have been misunderstood by subsequent parliaments; that laws have been enacted professedly in conformity to this statute, but entirely subversive of it; and that anomalies have crept into the weights and measures of England, and of this Union, which it appears to be impossible to trace to any other source.

The only notice which most of the modern writers upon English weights and measures have taken of this statute has been, to censure it for taking kernels of wheat as the natural standard of weights; with the very obvious remark that the wheat of different seasons and of different fields, and often even of the same field and the same season, is different. But the statute is chargeable with no such uncertainty. The statute merely describes how the standard measure of the exchequer, by the consent of the whole realm of England, was made. The article, for which of all others the measure was most wanted, was wheat; and a measure was wanted which should give it, as far as was practicable, in number, weight, and measure. It took, therefore, thirty-two kernels of *average* wheat from the middle of the ear, and found them equal in weight to the silver penny sterling, new from the mint, round and without clipping. It then drops the numeration of wheat; but proceeds to declare that twenty such pence make an ounce, twelve ounces one pound, and eight pounds a gallon of *wine*, and eight *gallons of wine* a London bushel, which is the eighth part of a quarter. It must be observed here, that it was not the *measure* but the *weight* of wine, which was used to form the standard bushel. It was not eight wine gallons, but eight gallons of wine. The bushel, therefore, filled with wheat, was a measure which, in the scales, would exactly balance a keg containing eight gallons of wine, deducting the tare of both the vessels. Now, the eighth part of this bushel, or the ale gallon, would be a vessel, not of the same cubic contents as the wine gallon, but of the same proportion to it as the weight of wheat bears to the weight of wine; the proportion between the commercial and nummulary weights of the Greeks; the proportion between our avoirdupois and troy pounds.

But neither the present avoirdupois, nor troy weights, were then the standard weights of England. The key-stone to the whole fabric of the system of 1266 was the *weight* of the silver penny *sterling*,

This penny was the two hundred and fortieth part of the tower pound; the sterling or easterling pound which had been used at the mint for centuries before the conquest, and which continued to be used for the coinage of money till the eighteenth year of Henry the Eighth, 1527, when the troy pound was substituted in its stead. The tower or easterling pound weighed three quarters of an ounce troy less than the troy pound, and was consequently in the proportion to it of 15 to 16. Its penny, or two hundred and fortieth part, weighed, therefore, 22½ grains troy; and that was the weight of the thirty-two kernels of wheat from the middle of the ear, which, according to the statute of 1266, had been taken to form the standard measure of wheat for the whole realm of England. It is also to be remembered, that the eight twelve ounce pounds of wheat, which made the gallon of wine, produced a measure which contained nearly ten of the same pounds of wine. The commercial pound, by which wine and most other articles were weighed, was then of fifteen ounces. This is apparent from the treatise of weights and measures of 1304, which repeats the composition of measures declared in the statute of 1266, with a variation of expressions, entirely decisive of its meaning. It says that, "by the " ordnance of the whole realm of England, the measure of the king " was made, that is to say: that the penny called sterling, round, and " without clipping, shall weigh thirty-two grains of wheat in the " middle of the ear. And the ounce shall weigh twenty pence; and " twelve ounces make the *London* pound; and eight pounds of *wheat* " make a gallon; and eight gallons make the London bushel." It then proceeds to enumerate a multitude of other articles, sold by weight or by numbers, such as lead, wool, cheese, spices, hides, and various kinds of fish; and, after mentioning nominal hundreds, consisting of 108 and 120, finally adds, "it is to be known that every " pound of money and of medicines consists only of twenty shillings " weight; but the pound of *all* other things consists of twenty-five " shillings. The ounce of medicines consists of twenty pence, and " the pound contains twelve ounces; but, in other things, the pound " contains fifteen ounces, and, in both cases, the ounce is of the weight " of twenty pence."

Wine and wheat therefore were both among the articles of which the pound consisted of fifteen ounces. By the statute of 1266, the gallon of wine contained eight such pounds of wine. By the statute of 1304, the gallon [for ale] contained eight such pounds of wheat; and the weight of wine contained in eight such wine gallons, and the weight of wheat contained in eight such corn or ale gallons, was equally the measure of the bushel.

The wine, to which the statute of 1266, and many subsequent English statutes exclusively refer, was the wine of Gascoign, a province at that, and for a long period, under the dominion of the English kings, the same sort of wine which now goes under the denomination of Claret, or Bordeaux. Its specific gravity is to that of distilled water as 9,935 to 10,000, and its weight is of 250 grains troy weight to the cubic inch.

ON WEIGHTS AND MEASURES. 27

With these data, we are enabled, accurately, to ascertain the dimensions and contents of the bushel, the ale gallon, and wine gallon, of 1266. The silver penny, called the sterling, to which 32 kernels of wheat were equiponderant, was equal to $22\frac{1}{2}$ grains troy. Its pound of twelve ounces was equivalent to 5400 grains troy. The pound of fifteen ounces, by which wheat and wine were weighed, was equal to 6750 grains troy. Eight such pounds were equal to 54,000 grains troy, which divided by 250, the number of grains troy, weighed by a cubic inch of Bordeaux wine, gives a wine gallon of 216 cubic inches.

There is no standard wine gallon of that age extant in England; but the weights and measures of England were established by law in Ireland as early as the year 1351: and by the act called Poyning's law, of 10 Henry VII., (1493,) all the then existing statutes of England, relating to weights and measures, were made applicable to Ireland. The changes since effected in England have not extended to Ireland; at least in relation to the measure of wine. The standard Irish wine gallon at this day is of 217.6 cubic inches; a difference almost imperceptible in the quantity of the gallon, from the legal standard of 1266, and the cause of which must have been this.

There was another law, probably of date more ancient than the year 1266, in which the measure of the wine gallon was fixed by a different process.

A statute of the year 1423, the second of Henry the Sixth, ch. ii, declares that, " *in old time* it was ordained, and lawfully used, that " tuns, pipes, tertians, hogsheads of Gascoigne wine, barrels of her- " ring and of eels, and butts of salmon, coming by way of merchan- " dise into the land, out of strange countries, and also made in the " same land, should be of certain measure; that is to say: the tun of " wine 252 gallons, the pipe 126 gallons, the tertian 84 gallons, the " hogshead 63 gallons, the barrel of herring and of eels 30 gallons, " fully packed, the butt of salmon 84 gallons, fully packed, &c.; but " that of late, by device and subtlety, such vessels have been of much " less measure, to the great deceit and loss of the king and his people, " whereof special remedy was prayed in the parliament." It then proceeds to re-enact that no man shall make in England vessels for those purposes, or bring wine, &c. into England in vessels of other dimensions, than those thus prescribed, upon penalty of forfeiture.

The ordinance of *old time,* referred to in this act, is not now among the statutes at large, and is therefore probably of more ancient date than the Magna Charta of 1225. As it regulated the size of casks, which, in the nature of the thing, were to be made in the country whence the wine was imported, it seems likely to have originated when Gascoign was under English dominion, and when the law of Bordeaux could be accommodated to the assize of the English ton. This assize of the ton is in its nature connected with the trade of the cooper, with the assize of hoops and staves, with the art of the shipbuilder, and with the whole science of hydraulics and of navigation. The measure and form of the ton must be accommodated to the character of the substance which it is to contain, and to the convenience and safety of its conveyance by sea. It must be adapted for stowage to the necessary form of the ship; to the volatile property of fluids; to the concussions of tempestuous elements. It is in the composition of the ton that the natural connection between the weight of water, and cubic linear measure, first presents itself. The burden of the ship is the weight of tonnage which it can bear afloat upon the waves; that weight is equal to the weight of water which it displaces; the measure of the ship must be taken by the builder in linear measure. Now eighty of the old easterling tower pounds make 432,000 grains troy weight, which, divided again by 250, the number of troy grains to a cubic inch of Bordeaux wine, give 1728 cubic inches, precisely the dimensions of an English cubic foot, one eighth part of which makes again the gallon of 216 cubic inches. And here we discover, again, the quadrantal or amphora of the Romans, the cubic foot containing 80 pounds of wine.

That the assize of the ton, which in 1423 was *of old time*, was equally well known and established in 1353, appears from a statute of that date, 27 Edward III., ch. 8, directing that all wines, red and white, should be gauged by the king's gaugers, and that in case less should be found in the tun or pipe than ought to be of right, *after the assize of the tun*, the value of as much as lacked should be allowed and deducted in payment.

The casks of Bordeaux wine were then and still are made for stowage in such manner that four hogsheads occupy one ton of shipping. The ton was of thirty-two cubic feet by measure, and of 2016 English pounds, of fifteen ounces to the pound, in weight; equal to 2560 of the easterling tower pound.

In comparing together the wine gallon as prescribed by the statute of 1266 and that derived from the assize of the tun, we find the former in the ascending ratio, beginning with the kernel of wheat and multiplying: the latter is formed in the descending ratio, beginning at the tun and dividing. In one process, the gallon is formed by weight; in the other, by measure. The hogshead of wine was the measure corresponding to the quarter of wheat: but there was a difference of eight pounds in their weight. The hogshead of wine weighed 504 and the quarter of wheat 512 pounds, of 15 ounces. The wine gallon of 216 cubic inches, prescribed by the statute of 1266, was thus an exact eighth part of the English cubic foot of 1728 inches.

The wine gallon therefore is the *congius* of the Romans, weighing ten nummulary and eight commercial pounds, and measuring exactly the eighth part of a cubic foot.

But the gallon of 216 cubic inches, the eighth part of the cubic foot, was derived originally from a measure of *water*, and was an aliquot part of the ton of shipping. The wine gallon of 1266 was made of eight easterling pounds of wheat; and, therefore, contained of water eight corresponding commercial pounds. But if the gallon of water, weighing eight pounds, was of 216 solid inches, the gallon of Gascoign wine, to be of the same weight, would be of 217.6 solid inches, the precise contents of the standard Irish gallon to this day: and the specific gravity of that wine being to that of wheat as 143 to 175, the corn gallon, balanced by this Irish gallon of 217.6 inches, must be of 266.17 cubic inches. The Rumford corn gallon of the year 1228, examined by the committee of the House of Commons in 1758, was found to be of 266.25 cubic inches. The Irish wine gallon and the Rumford corn gallon of 1228 were both made, with an accuracy which all the refinements of art of the present age could scarcely surpass, from the standard measure made, as the statute of 1266 declares, by the consent of the whole realm, and precisely in the manner therein described.

But, as the hogshead, measuring eight cubic feet, was required by the assize of the tun to contain only sixty-three gallons of wine, it followed of course that the gallon thus composed was of 219.43 cubic inches; and as the weight of eight such gallons of wine was to form the bushel, the proportion of the weight of wine being to that of wheat as 143 to 175, the bushel would be of 2148.25 cubic inches, which is within two inches of the Winchester bushel.

This system of weights and measures has been, by many of the modern English writers on the subject, supposed to have been *established* by the statute of 1266. But, upon the face of the statute itself, it is a mere exemplification of ancient ordinances. The coincidences in its composition with those of the ancient Romans, proved by the letter of the Silian law, and by the still existing congius of Vespasian; with those of the Greeks, as described by Galen, and as shown by the proportions between their scale weight and their metrical weight; and with that of the Hebrews, as described in the prophecy of Ezekiel; show that its origin is traceable to Egypt and Babylon, and there vanishes in the darkness of antiquity. As founded upon the identity of nummulary weights and silver coins, and upon the relative proportion between the gravity and extension of the first articles of human traffic, corn and wine, it is supposed to have originated in the nature and relations of social man, and of things.

It has been said, that the first inroad upon this system in England was made by Edward the First himself, by destroying the identity between the money weight and the silver coin. From the time of the Norman Conquest, and long before, that is, for a space of more than three centuries, the tower easterling or sterling pound had been coined into twenty shillings, or two hundred and forty of those silver *pennies*, each of which weighed thirty-two kernels of wheat from the middle of the ear. Edward the First, in the year 1328, coined the same pound into two hundred and forty-three pennies of the same standard alloy. From the moment of that coinage, the penny called a sterling, however round, however unclipped, had lost the *sterling* weight, though it still retained the name. This debasement of the coin, once commenced, was repeated by successive sovereigns, till, in the reign of Edward the Third, the pound was coined into twenty-five shillings, or 300 pennies. The silver penny then weighed only $25\frac{2}{5}$ kernels of that wheat of which the penny of 1266 weighed 32. It is probable that, in reducing the weight of their coins, none of those sovereigns were aware that they were taking away the standard of all the weights and of all the vessels of measure, liquid and dry, throughout the kingdom: but so it was. It destroyed all the symmetry of the system. It has been further affected by the introduction of the troy and avoirdupois weights.

The standard measures of the exchequer had been made by the rules set forth in the statutes of 1266 and 1304. These standards were kept in the royal exchequer. In process of time, the standards themselves fell into decay, and called for renovation. In the year 1494, shortly after the termination of the long and sanguinary wars between the houses of York and Lancaster, Henry the Seventh, in the tenth year of his reign, undertook to furnish forty-three of the principal cities of the kingdom with new copies of all the standard weights and measures then in the exchequer. They were accordingly made and delivered to the representatives in parliament of the respective counties: but it was soon discovered that they were all defective, and not made according to the laws of the land. From what cause this had arisen, does not appear; but that the laws of the land to which they referred, namely, the statutes of 1266 and 1304, were and continued to be entirely misunderstood, is abundantly apparent from the statute which was made the very next session of parliament, 1496, to remedy the evil.

This act, after reciting the extraordinary attention of the king in having made at his great charge and cost, and having distributed, all those county standards of weights and measures, according to the old standards in the treasury; and after stating the disappointment which had ensued, upon the discovery of more diligent examination that they were all defective and not made according to the old laws and statutes, proceeds to ordain, that the measure of a bushel contain eight gallons of wheat, that every gallon contain eight pounds of wheat, *troy* weight, and every pound contain twelve ounces of troy weight, and every ounce contain twenty *sterlings*, and every sterling be of the weight of thirty-two corns of wheat that grew in the midst of the ear of wheat, *according to the old laws of the land:* and the new standard gallon, after the said assize, was to be made to remain in the king's treasury forever. All the weights and measures, which had been sent by the act of the former year throughout England, were directed to be returned: others, conformable to the new standard, were to be made from them and sent back, after which, all weights and measures were to be made conformably to them.

It is from the terms of this statute, that many of the English writers have concluded that the kernel of wheat was the original standard of English weights. It is by this statute made the standard of troy weight; but it was not so according to the *old laws of the land*. It was not so in the measure declared in 1266, and 1304, to have been made by the consent of the whole realm of England. To prove this, it is only necessary to compare the statutes together.

The two first declare, that an English *penny, called a sterling,* round and without any clipping, will weigh thirty-two corns of wheat from the midst of the ear. That penny was the two hundred and fortieth part of the old tower pound, and was one-sixteenth lighter than troy weight. The weight of that penny in 1266 is therefore now known, but appears not to have been known to the parliament of 1496. For the tower pound was then coined into thirty-seven shillings and six pence sterling, and, consequently, the penny called a sterling, instead of *then* weighing thirty-two grains of the wheat, which it weighed in 1304, would have weighed only seventeen of the same grains.

The term *penny*, therefore, is dropped in the act of 1496, but the term sterling is retained, and improperly applied to the penny *weight* troy. The penny of 1266 was both weight and coin. In 1496, the penny had ceased to be a coin, and the penny sterling, which was yet money, weighed little more than half what it had weighed till after 1304. The penny weight troy was never called *a sterling*, any where, or at any time, but in this act of 1496. It was neither the weight of the old tower standard, nor was it the penny sterling of Henry the Seventh's own coinage.

The statute of 1496 inverts the order of the old statutes; it is not a *composition*, but an *analysis*, of measures. It begins with the bushel, and descends to the kernel. The act of 1266, to make the weight, number, and measure, of corn, money, and wine, begins with the kernel, and ascends by steps to the weight of coin; thence, to the measure of wine, by the weight of corn; thence, to the measure of corn, by the weight of wine. The mere process of the composition establishes the proportional measures. The statute of 1496 destroys the proportion altogether. It says that every gallon shall contain eight pounds of wheat *troy weight,* and every pound twelve ounces of troy weight. It substitutes, therefore, instead of the *weight* of the gallon of wine, prescribed by the statute of 1266, the *measure* of the wine gallon, for the eighth part of the bushel. The gallon, established by this act of 1496, is the gallon of two hundred and twenty-four cubic inches; the Guildhall gallon, which in 1688 was found by the commissioners of the excise to be of that capacity. It contains eight pounds troy weight of wheat, and, consequently, eight pounds avoirdupois of Bordeaux wine, of 250 grains troy to the cubic inch. Its bushel would contain seventeen hundred and ninety-two cubic inches; but if such a bushel ever was made, as the act required, it never was used as a standard. It must have been found to fall too far short of the old standards still existing; and the real standard bushels of Henry the Seventh, in the exchequer, instead of being made according to the process prescribed in his law of 1496, must have been copied from the older standard bushels then existing.

The gallon of two hundred and thirty-one inches was also a gallon made under the statute of 1496. But the wheat is of that kind thirty-two grains of which equipoise the penny of the old tower pound; while the wheat that forms the gallon of two hundred and twenty-four inches, is that of which thirty-two kernels weigh a penny weight troy. The weight of the corn in both gallons would be the same; but that, of which each kernel upon the average would be one sixteenth heavier than those of the other, would, by the combined proportion of gravity and numbers, occupy one thirty-second less of space. This is precisely the difference between the gallons of two hundred and twenty-four and two hundred and thirty-one solid inches.

The debasement of the coin had destroyed its original identity with the money weight. The substitution of *troy* weight, instead of the old easterling pound, for the composition of the gallon, destroyed the coincidence between the water gallon, derived from the ton, the eighth part of the cubic foot, and the wine gallon, containing eight money pounds of wheat. The wine gallon of two hundred and twenty-four, or two hundred and thirty-one cubic inches, no longer bore the same proportion to the cubic foot of water; one consequence of which was, that the hogshead of Bordeaux wine, which the law required to contain sixty-three gallons, no longer contained that number of English gallons; but, from that day to this, has contained from fifty-nine to sixty-one. It still contains at least sixty-three Irish gallons.

The act of 12 Henry VII, (1496,) intended, upon the face of it, to be a mere repetition of the statutes of 1266, and 1304, was thus a total subversion of them. It was founded upon two mistakes; the first, a supposition that the penny sterling, described in those statutes, was the penny weight troy; and the second, a belief that it was the *measure*, and not the *weight*, of eight gallons of wine, which constituted the bushel. The causes of these mistakes were, first, that, in the lapse of two centuries, a great portion of which had been a period of calamity and civil war, the successive debasements of the coin had reduced the penny sterling to about half its weight as it was when made the standard of comparison with thirty-two kernels of wheat; and finding that the penny sterling of their own time, if used to make the new standard bushel, would reduce its size by nearly one half, which had perhaps been the mistake upon which the act of 1494 had been made, they must hastily have concluded that it was the penny weight troy, which was intended by the old statutes. The second was a misapplication of the term *gallon*, which, in its original meaning, and in its popular signification to this day, is exclusively a measure of liquids, and not of dry substances. In the statute of 1266, it is expressly called the gallon of *wine*. In the act of 1304, it was called the gallon, without addition, but meaning the same wine gallon. The measure for corn was the bushel; and its subdivisions were the peck, pottle, and pint. The eighth part of the *measure* of a bushel was first called a *gallon*, because it was used as the measure of the ordinary liquids brewed from grain, beer and ale. There never was properly any corn gallon; and the term was misapplied even to denote the measure of beer. Being a vessel of different dimensions, it ought to have had a different name; and that alone would have prevented them from ever being mistaken the one for the other.

The parliament of 1496 were seduced by those expressions, so often re-echoed from year to year, and from century to century, that there should be but one weight, and one measure, throughout the land. They mistook the uniformity of proportion, for the uniformity of identity. They construed the threefold "one measure of wine, and one measure of ale, and one measure of corn," ordained in Magna Charta, as if it meant that those three one measures should be the same. That these mistakes should be made, is not surprising when we consider that, in 1496, the art of printing was but in the cradle; that no collection of the statutes had ever been printed; that the languages in which the statutes of 1266, and 1304, had been enacted, the Norman French, and the barbarous Latin of the thirteenth century, were no longer in use, at least in parliament; that the very records, by which the weight of the penny sterling in 1266 might have been ascertained, were, perhaps, not known to exist. But it is not so easy to explain how they could mistake the penny of the old easterling pound, which was still, and continued for forty years after to be, used at the mint for coining money, for the penny troy weight.

We have seen that, in 1304, the easterling pound of twelve ounces was the money pound, and that the corresponding commercial weight was a pound of fifteen of the same ounces. These were the result of a rough and inaccurately settled proportion between the specific gravity of wheat and wine, or wheat and water. Mr. Jefferson has justly remarked, that the difference between the specific gravity of wine and of water is so small, that it may safely, as between buyer and seller, be disregarded. And it was disregarded by those two acts of parliament, one of which made the wine gallon an eighth part of the cubic foot of water, while the other made it equiponderant to eight easterling pounds of wheat. So the proportion of the two pounds of twelve and fifteen of the same ounces was, upon a rough estimate, that the proportional weight of wheat and wine was as four to five, or as fourteen to seventeen and a half; and it was afterwards assumed as of fourteen to seventeen. But if trade, and even legislation, may safely neglect small quantities, nature is no such accountant of more or less. It has been shown that the water gallon, of eight easterling twelve ounce pounds of wheat, corresponds with the gallon of two hundred and sixteen cubic inches, the eighth part of the cubic foot; but that, when a Bordeaux *wine* gallon is to be made, containing the same eight pounds of wheat, it produces a gallon, not of 216, but of 217.6 cubic inches. In the mode of forming the gallon and bushel, described in the act of 1266, it is not the loose calculations of man, but the unerring hand of nature, that establishes the proportions. The vessel that would hold eight money pounds of wheat was the wine gallon. The vessel that would balance, filled with wheat, eight gallons of that wine, was the bushel. Then, if a half peck of this bushel was taken for a beer gallon, its proportion to the wine gallon would not be of fifteen to twelve, nor of seventeen to fourteen, but of one hundred and seventy-five to one hundred and forty-three.

When the avoirdupois or the troy weights were first introduced into England, has been a subject of controversy among the English writers, and is not ascertained. The names of both indicate a French origin: but that no new weight or measure was brought in by the Norman Conqueror is certain, and the statute of weights and measures of 1304 proves that neither of them were *then* recognized by law. One of the most learned writers upon the coins,* says that troy weights were first established by this statute of Henry VII. of 1496; that it was owing to the *intercursus magnus*, or great treaty of commerce concluded between England and Flanders the year before; that the Flemish pound was adopted as a compliment to the Duchess of Burgundy, and for the mutual convenience of all their payments, which would then be adjusted by the same pound.

This conjecture is ingenious, but not well founded. The statute of 1496 did, in fact, legitimate troy weight for the composition of the gallon and the bushel, but it professed, and intended to introduce, no new weight or measure. Its purpose was to re-enact the composition of weights and measures of 1266. It was a legislative error, intended to correct another error committed at the last preceding session of parliament in 1494, before the intercursus magnus was concluded. Instead of correcting the error, it rendered it irreparable. It was so far from correcting the error, that, although a standard wine gallon was made under this statute, which was the Guildhall wine gallon of two hundred and twenty-four inches, there never was a standard bushel made by the rule prescribed in this statute: and if there had been, its cubic contents would have been not one inch more or less than seventeen hundred and ninety-two.

The troy weight was never used by Henry VII. at the mint at all. He made a wine gallon by it, because the difference between a gallon of 217.6 inches, which was the old standard, and one of 224, which was made by his troy weights, was not large enough to make its incorrectness apparent. It was scarcely the difference of a small wine glass upon a gallon: and, as it was a difference of excess over the contents of the old standard, it might naturally be attributed to the decays or inaccuracy of that. He ordained that a bushel should be made by it; but a bushel made from the *measure* of his wine gallon, a bushel of seventeen hundred and ninety-two inches, would have contained at least three hundred and thirty inches, nearly twelve pounds in weight less than any of the old standards. This would have been found a difference utterly intolerable. It would have been necessary to recal and break up the new standards a second time, and to acknowledge a second error greater than the first. So the statute, so far as related to the composition of the bushel, was suffered to slumber upon the rolls; the old standard bushels were still retained; and new ones were also made, not by the troy, but by

---

* Clarke.

the avoirdupois pound of wheat: and hence it is that standard bushels of Henry the Seventh exist at the exchequer, one of 2124 inches, which is the old standard, and one of 2146 inches, which is the Winchester bushel, and, at the same time, corn gallons of 272 inches.

That the troy weight was not introduced into England by Henry the Seventh is further proved by two statutes; one of 1414, 2 Henry V. ch. 4, and one of 1423, 2 Henry VI. st. 2, ch. 4: in the first of which it is ordained, that the goldsmiths should give no silver worse than of the allay of the English sterling, and that they take for a pound of *troy* gilt but forty-six shillings and eight pence at the most; and, in the second, that silver, not coined, in plate, piece, or in mass, being of as good allay as the sterling, should not be sold for more than thirty shillings the pound troy, besides the fashion, because the same was of no more value at the coin than thirty-two shillings. The tower easterling pound was at that time coined into thirty shillings, and the value of the troy pound of the same alloy was, of course, thirty-two.

From these two statutes it is apparent that, nearly a century before Henry the Seventh, the troy pound was used by the goldsmiths, who were the bankers of that age, and were foreigners, for weighing bullion and plate; and that the proportion of the troy pound to the tower money pound was as sixteen to fifteen.

That the troy pound, though adopted by Henry the Seventh for the composition of the bushel and gallon, was not introduced by him at the mint, appears equally clear. About the middle of the last century, Martin Folkes published his tables of English coins, in which he cited a verdict remaining in the exchequer, dated 30th October, 1527, 18 Henry VIII. in which are the following words: "And "whereas, heretofore, the merchant paid for coynage of every *pound* "*towre* of fyne gold, weighing xi. oz. quarter troye, 2*s*. 6*d*. "Nowe it is determined by the King's highness, and his said coun- "celle, that the aforesaid ponnde towre shall be no more used and "occupied; but all manner of golde and sylver shall be wayed by the "pounde troye, which maketh xii. oz. troye, which excedith the "pounde towre in weight three quarters of the ounce." [*Clarke*, p. 15.]

A French record of much earlier date, from the register of the Chamber of Accounts at Paris, cited also by Folkes, shows that, early in the fourteenth century, there were among the weights in common use in France two marks, of different gravity, one of troy, and the other of *Rochelle*, in the same proportion to each other, and that the last was called the mark *of England*.

The Rochelle and easterling pound was therefore the same; and that was the pound, eight of which in spring water were contained in the eighth part of the cubic foot, and formed the gallon of 216 cubic inches.

This proportion, as has been observed, was totally destroyed by the substitution of the troy pound by Henry the Seventh, in 1496, instead of the Rochelle pound, for the composition of the bushel and gallon.

As, by the treatise of weights and measures of 1304, only two weights are mentioned, by which it asserts that all things were weighed, this towre pound of twelve ounces, and the corresponding commercial pound of fifteen of the same ounces, it is clear that the troy weight was then unknown, or at least not used in England. But this reign of Edward the First was also the period when the foreign commerce of England began to flourish. In 1296, the famous mercantile society, called the Merchant Adventurers, had its first origin; and another society of natives of Lombardy, the great merchants of that age, about the same time established themselves in England, under the protection of a special charter of privileges from Edward. These Lombards soon became the goldsmiths and bankers in England. [*Hume*, vol. ii. p. 330, ch. 13.] In the year 1354, the balance of exports above the imports was of more than 250,000 pounds; and as the balances of that age could be paid for only in specie, whenever the balance was in favor of England it must have brought much foreign money into the kingdom. The pound of the goldsmiths and bankers was the troy weight, and by them, there can be little doubt, it was first introduced. The pound of fifteen ounces troy must have been introduced at the same time, by an accommodation of that weight to the old English rule—that when bullion and drugs were weighed by a pound of twelve ounces, all other things were weighed by a pound of fifteen of the same ounces. This pound of fifteen ounces, or 7200 grains troy, has never been recognized in England by law; but, in many parts of England, it has been used under the name of the merchant's weight: and eight such pounds of wheat form precisely the gallon of 280 cubic inches, of which the standard quart in the exchequer is the fourth part.

The time and occasion of the introduction of the avoirdupois pound into England is no better known than that of the troy weight. But it may be inferred, from the ancient statutes, that it was brought in by the same foreign merchants, with the troy, and as the corresponding weight to that as the weight for bullion and pharmacy. The first time that the term avoirdupois is used in the English statute book, is in the 9th of Edward III. stat. 1, ch. i. (1335,) the very statute which authorizes merchant *strangers* to buy or sell corn, wine, *avoirdupois*, flesh, fish, and all other provisions, and victuals, wool, cloths, wares, merchandises, and all other vendible articles in any part of England.

Eighteen years afterwards, in the celebrated statute staple of 1353, [27 Edward III. ch. 10.] is the following provision: "Foras-
" much as we have heard that some merchants purchase *avoirdupois*,
" woollens, and other merchandises, by one weight and sell by ano-
" ther, and also make deceivable diminutions upon the weight, and
" also use false measures and yards, to the great deceit of us and of
" all the commonalty and of honest merchants: We therefore will
" and establish, that one weight, one measure, and one yard, be

"throughout the land, as well without the staple as within, and that "woollens, and all manner of *avoirdupois*, shall be weighed by balances," &c.

In these two statutes, the term avoirdupois manifestly refers, not to the weight, but to the article weighed. It means all *weighable* articles, in contradistinction to articles sold by measure or by tale; and the *one weight*, meant by the statute staple, is the easterling pound of fifteen ounces, mentioned in the statute of weights and measures of 1304. Money and bullion were not included among these *weighable* articles, because they had a special weight of their own, and because money was current by *tale*. Grain and liquids were not weighable articles, because they were bought and sold by *measure*. Hence arose naturally the practice of calling all articles bought and sold by weight in the traffic of these merchant strangers, articles *having weight*, or weighable.

That this is the meaning of the term avoirdupois is also demonstrated by an act of 1429, (8 Henry VI. ch. 5.) which reciting these regulations of Edward the Third, expressly says, that they require woollens, and all manner of *weighable* things, [toutz manerz des choses poisablez] bought or sold, to be weighed by even balance, with weights sealed according to the standard of the exchequer.

The terms *avoirdupois*, and *choses poisablez*, were therefore synonymous. But the merchant strangers had a weight of their own; the corresponding commercial weight proportional to their pound troy. This was the weight now called the avoirdupois pound, of sixteen ounces, but the ounce of which was not the same as that of the troy weight. The standard easterling pound of fifteen ounces at the exchequer weighed 6750 grains troy. The avoirdupois pound of the merchant strangers weighed 7000. The difference between them was but of about half an ounce, and one sees instantly what temptations and opportunities this slight difference furnished to those unprincipled merchants of whom the statute staple complains, of buying by their own foreign larger weight, and selling by the weight of the exchequer.

The statute staple of 1353, and the act of 2 Henry VI., 1423, are both evidences of the conflict between the mint and exchequer easterling pounds on one side, and the troy and avoirdupois weights of the merchant strangers on the other. In this struggle, the latter ultimately prevailed, and completely rooted out the old English weights. The troy weight, being adopted by Henry the Seventh, in 1496, for the composition of the bushel and gallon, and by Henry the Eighth in 1527 for making money at the mint, the avoirdupois pound came in as the corresponding commercial pound; and a statute of 24 Henry VIII., ch. 3, 1532, directs, that beef, pork, mutton, and veal, shall be sold by weight, *called* haverdupois; the very use of which expression, *called* haverdupois, indicates that it was a denomination, as applied to the weight, of recent origin, and that the weight itself had not been long in general use for any purpose.

A statute of the preceding year, 23 Henry VIII., ch. 4, 1531, s. 13, had directed, that every cooper should make every barrel for ale " *according to the assize specified in the treatise called Compositio Men-* " *surarum;* [the statute of 1304] that is to say, every barrel for ale, " shall contain thirty-two gallons of the said assize, or above, of the " which eight gallons make the common bushel to be used in this " realm of England," &c.

By this statute, the ale gallon was expressly declared to be the eighth part of the *measure* of the bushel. Now, it has been proved, that, by the *Compositio Mensurarum,* the bushel was a measure containing of wheat the *weight* of eight gallons of wine. The eighth part of this *measure*, therefore, being the ale gallon, *must* bear the same proportion to the wine gallon, as the specific gravity of wheat bears to that of wine: and the wine gallon of 231 inches having been made by the rule of the *Compositio Mensurarum,* but by the troy weight of the statute of 1496, that is to say, weighing eight troy pounds of the wheat thirty-two kernels of which were equiponderant to the penny sterling of 1266, the bushel, to balance eight such gallons of wine, must of necessity contain sixty-four avoirdupois pounds of wheat, and measure 2256 cubic inches. The eighth part of this measure is the gallon of 282 inches, which is to this day the standard ale and beer gallon of the British exchequer, and of these United States.

And thus we have seen that all the varieties of standard gallons and bushels, which have been found, from the Irish gallon of 217.6 cubic inches, to the ale gallon of 282, and from the ordained, but never made, bushel of 1792 inches, prescribed by Henry the Seventh's act of 1496, to the bushel of 2256 inches, of which the ale gallon is the eighth part, are distinctly traceable to the inconsistency of human laws, and the consistency of the laws of nature.

The statute of 1496 changed the contents of both the gallons, and of the bushel, without intending it. For, although the bushel of 1792 inches was never made, or at least never deposited as a standard at the exchequer, yet new standard bushels were made from the new wine gallon, by the rule of the Compositio Mensurarum, and they produced the bushel of 2224 of Henry the Seventh, still extant at the exchequer.

The *Winchester* bushel of the exchequer, however, was not thu made. It was found in the year 1696 to contain 2145.6 cubic inches of spring water. Its ale gallon, therefore, by the statute of 1531, and the Compositio Mensurarum, would have been of 268.2, and its wine gallon of 219.2 cubic inches. Its difference from the proportions of the Irish wine gallon, and the Rumford corn gallon, is so slight, that there can be no doubt it was copied from a model made by the statute of 1266.

That the capacity of the wine gallon, although it was very essentially changed, was not intended or understood to be changed by the statute of 1496, is proved by the statute of 28 Henry VIII. ch. 14, 1536; which re-enacts the old statutes requiring that the tun of wine should contain 252 gallons, and all other casks of the same proportion, including the hogshead of sixty-three gallons. Now, the hogshead which contained sixty-three gallons of 217.6 cubic inches, could contain no more than sixty-one and a quarter gallons of 224 inches, nor more than fifty-nine and one third gallons of 231 inches. The ordinary Bordeaux hogshead contains from fifty-nine to sixty gallons: and the size of this cask, being formed of a certain number of staves of settled dimensions, and made by the coopers in particular forms, has passed down, from age to age, without alteration; while the laws of England, and those of several of the United States, have required that it should contain sixty-three gallons of 231 cubic inches, because, five hundred years ago, the laws required it to contain sixty-three gallons of 219.5 inches.

In the reign of Elizabeth, the change, which had been effected in the wine gallon by the act of 1496, was discovered in its effects upon another branch of trade; but the cause of the change appears not to have been perceived. The statute of 13 Elizabeth, ch. 11, (1570,) recites, that the people employed in the herring fishery " had, time " out of mind, used to pack their herring in barrels containing *about* " thirty-two gallons of usual *wine* measure, which assize had always " beeen gauged and allowed in the city of London, yet the measure " had lately been quarrelled at by certain informers, for not contain- " ing thirty-two gallons by the old measure of standard, *which they* " *never did*, though peradventure *the extremity of old statutes in words,* " *by some men's construction*, might be stretched to require so much." It then enacts, " that thirty-two gallons wine measure, which is *about* " twenty eight gallons by old standard, shall be the lawful assize of " herring barrels, any old statute to the contrary notwithstanding." This was cutting the Gordian knot. The wine gallon here referred to was the gallon of 231 inches, made by the troy weight wheat of Henry the Seventh. The *old standard* is the corn gallon of 1266. which, according to the Rumford quart examined by the committee of the House of Commons in 1758, was of 264.8, or, according to the Rumford gallon, was of 266.25 cubic inches. Now twenty-eight gallons of 264 inches are of precisely the same capacity as thirty-two gallons of 231. But, as the wine gallon at Guildhall, though it showed 231 inches by the gauge, did, in fact, contain seven inches less, and as the herring barrels were gauged according to the Guildhall gallon, they would have fallen short nearly one gallon in the barrel of twenty-eight gallons by the *old standard:* and the act, the object of which was to rescue the herring fishers from the fangs of informers, is cautious not to tie them down to too close a measure. The old statutes, the construction of which the act professes to consider as doubtful, are not named; but the act of 23 Henry VIII., 1531, *must* have been that upon which the informers had quarrelled at the assize of the barrels used by the herring fishers.

That act requires that the coopers should make barrels of thirty-two gallons for ale, according to the assize of the Compositio Mensurarum—gallons, eight of which make the common bushel. Now, the act of 1496 had expressly directed that *every* gallon should contain eight pounds of wheat *troy* weight, and that the bushel should contain eight such gallons of wheat. But this law, so far as it prescribed a new bushel, had never been executed: the old standard bushels remained. So that the statute for the coopers of 1531 was on the side of the informers, and the statute of weights and measures of 1496 was on the side of the herring fishers. Parliament found no expedient for the difficulty but to declare the usual existing size of the herring barrels lawful, and to set all the old statutes aside, with a non obstante.

If the parliament of 1496, contrary to their avowed intention, did actually change the capacity of the wine and corn gallons, and did ordain a much greater change of the capacity of the bushel, these varieties, effected by the law, while they were unknown to the legislators, were still less likely to come to the general knowledge of the people. The eagle eyes of informers would occasionally discover that the measures of the people fell short of the standards of the law: but the people took the standards as they came, and used the measures which they and their fore-fathers, time out of mind, had employed.

The Restoration of 1660, after the convulsions of a civil war, formed a new æra, not only in the political history of England, but in that of their vessels of capacity. It was then that a new system of taxation commenced by the excise upon liquors. About the same time, also, commenced a new æra in the philosophical and scientific pursuits of the English nation, by the establishment of the Royal Society. Both these events were destined to have an important influence upon the history of English weights and measures.

The excise was a duty levied, by the gallon, upon malt liquors, and upon wines. The malt liquors were measured by the standard gallon at the treasury, made according to the cooperage act of 1531, by the rule of the *Compositio Mensurarum*, applied to the troy weight wine gallon of the statute of 1496. This gallon, therefore, was neither the wine gallon of 1496, nor the eighth part of the old standard bushel, nor of the Winchester bushel, but the gallon which, if filled with wheat of the troy weight specified in the statute of 1496, would balance the wine gallon of 231 inches; it was, therefore, a gallon of 282 cubic inches. The wine was measured by the gauge of the wine gallon at Guildhall.

Taxation and philosophy now began to speculate, at the same time, upon the weights and measures of England. In 1685, the weight of a cubic foot of spring water was found, by an experiment made at Oxford, to be precisely 1,000 ounces avoirdupois; and, in 1696, the Winchester bushel was found to be of 2145.6 cubic inches, and to contain, also, 1,000 ounces avoirdupois weight of wheat. Yet so totally lost were all the traces of the old easterling pounds of twelve and fifteen ounces, that this coincidence between the cubic foot of water, and the 1,000 ounces avoirdupois, gave an erroneous direction to further inquiry; for the real original connection between the cubic foot and the English bushel was not formed by avoirdupois weights and water, but by the easterling pound of twelve and fifteen ounces and Gascoign wine. It was the principle of the quadrantal and congius of the Romans, applied to the foot and the nummulary pound of the Grecks; the measure which, by containing eight pounds of wheat, was intended to contain, at the same time, ten of the same pounds of wine.

In the year 1688, the commissioners of the excise instituted an inquiry why beer and ale were always gauged at 282 cubical inches for the gallon, and other exciseable liquors by the wine gallon of 231 inches. They addressed a memorial to the Lords of the Treasury, stating these facts, and that, being informed the true standard wine gallon ought to contain only 224 cubical inches, they had applied to the Auditor and Chamberlains of the *Exchequer*, who, upon examination of the standard measures in their custody, had found three standard gallons, one of Henry the Seventh, and two of 1601, which an able artist employed by them had found to contain each 272 cubical inches; that, finding no wine gallon at the Exchequer, they had applied to the Guildhall of the city of London, where they were informed the true standard of the wine gallon was, and they had found, by the said artist, that the same contained 224 cubical inches only: and they added, that the gallons of other parts of the kingdom used for wine, had been made and taken from the Guildhall gallon.

In consequence of this memorial, the Lords of the Treasury, the 21st May, 1688, directed an authority to be drawn for gauging according to the Guildhall gallon; which was accordingly done: but the authority does not appear to have been signed. The ale gallon at the Treasury was of 282 inches; but the order of the Lords of the Treasury for the benefit of the revenue would have reduced the gallon, both for malt liquors and wine, to the Guildhall gallon of 224. The merchants immediately took the alarm, and petitioned that they might be allowed to *sell* by the same gauge, of 224 inches to the gallon, by which they were to be required to pay the customs and excise. The commissioners of the customs not agreeing with those of the excise, on the proposal for a new gauge, the opinion of Sir Thomas Powis, the attorney general, was taken upon it, who advised against any departure from the usage of gauging, because the Guildhall gallon was not recognized as a legal standard, and because by any of those at the exchequer the king would *be vastly a loser*.

Sir Thomas Powis then refers to the old statutes, prescribing that eight pounds should make a gallon; and particularly to that of 1496, requiring that the eight pounds should be of wheat; *and as* there was to be *one measure throughout the kingdom*, which could not be, unless it was adjusted by some óne thing, and that seemed to be intended wheat, therefore *he did not know* how 231 cubical inches came to be taken up, but did not think it safe to depart from *the usage*, and therefore the proposal was dropped.

Sir Thomas Powis's *reasoning*, upon the statute of 1496, was perfectly correct. That statute, as well as many others, does ordain one measure throughout the kingdom; it does ordain that *every gallon* shall contain eight pounds troy weight of wheat of thirty-two kernels to the pennyweight troy, which it strangely calls a sterling. Sir Thomas did not know how 231 inches came to be taken up; because he did not know that the statute of 1496 had substituted the troy for the old easterling weight in the composition of the gallon. It was that change that brought up the 231 inches: for, if eight easterling twelve ounce pounds of wheat filled a gallon of 217.6 inches, eight troy pounds of the same wheat must of necessity fill a gallon of 231.

The Guildhall wine gallon contained also eight troy pounds of wheat; but it was wheat thirty-two kernels of which weighed a pennyweight troy. Every kernel on the average was $\frac{1}{17}$ heavier than that which had been used for the composition of the gallon and bushel of 1266. The average kernel being specifically heavier, a pound weight of it occupied less space: on the other hand, the corn of lighter kernel would require a greater number of kernels to make up the same weight. The gallon of 1496 was to contain 61,440 kernels, weighing in the aggregate eight pounds troy: and they would fill a space of 224 cubic inches. To make the same weight, eight pounds troy would take 65,280 kernels of the wheat of 1266: but these 65,280 kernels would fill a space of 231 cubic inches. The difference between the two was a compound of the increase of numbers and the diminution of weight.

The advice of Sir Thomas Powis was, however, followed without further inquiry, and the use of the gauging rods was continued. But in 1700, the same inconsistency of the statutes, which, in the reign of Elizabeth had bred the quarrel between the informers and the herring barrels, generated a lawsuit between commerce and revenue. It has been seen, that, by a statute of 2 Henry VI., ch. 11, confirmed by subsequent acts of 1483, 1 Richard III., ch. 13, and of 1536, 28 Henry VIII., ch. 14, it had been ordained that every *butt* or pipe of wine imported should contain 126 gallons. The original statutes had reference to the Gascoign or Bordeaux wines, the casks of which were proportioned to the ton of thirty-two cubic feet. When afterwards the importation of Spanish wines became frequent, they were brought in casks of different dimensions from the assize: and the statute of Richard the Third, reciting that their butts had theretofore often been of 140 or 132 gallons, and complaining that they had been of late fraudulently reduced to 120 gallons or less, prescribed that they should thenceforth be of at least 126 gallons. The old fashion, of 140 gallons or more to the *butt* of Malmsey and other Spanish wines, was then restored: and as the law was satisfied if the butts were of 126 gallons or more, their size beyond the usual dimensions of the Gascoign standard remained unnoticed till the fiscal officer became interested in their contents. When customs and excise came to call for their share of the Malmsey, the merchants for some years paid upon the butt as if it had contained only the 126 gallons required by the law. But this calculation could not long suit the revenue. An action was brought by the officers of the customs against Mr. Thomas Barker, an importing merchant, for the duties upon sixty butts of Alicant wine, for which he had paid, as if containing 126 gallons; but which, in fact, contained 150 gallons each. The crown officers showed, that the butt was to contain, by law, 126 gallons; and Mr. Leader, the city gauger, Mr. Flamstead, and other skilful gaugers, all agreed, that a wine gallon ought to contain 231 cubical inches, and no more; that there was such a gallon, kept from time out of mind at Guildhall, (they were in this mistaken, for it contained only 224 inches,) that the wine gallon was less than the corn gallon, which was of 272, and the ale gallon, which was of 282 cubical inches.

The defendant insisted that the laws had directed that a standard should be kept at the Treasury; that there was one there, containing 282 cubic inches; that by that measure he had paid the duty; that the Guildhall gallon was no legal standard: and merchants, masters of ships, and vintners, of twenty, thirty, forty years experience, all testified that Spanish wine always came in butts of 140 or 150 gallons or more. Whether Mr. Thomas Barker, when he came to *sell* his wine, retained his contempt for the Guildhall gallon, is not upon the record.

After a trial of five hours, the attorney general made it a drawn battle; agreed to withdraw a juror; and advised to leave the remedy to parliament: and this was the immediate occasion of the statute of 5 Anne, ch. 27. sec. 17., by which the capacity of the wine gallon is fixed, and has ever since remained, at 231 cubical inches. This act declares, that any round vessel, commonly called a cylinder, having an even bottom, and being seven inches diameter throughout, and six inches deep from the top of the inside to the bottom, or any vessel containing 231 cubical inches and no more, shall be deemed and taken to be a lawful wine gallon: and it is hereby declared, that 252 gallons, consisting each of 231 cubical inches, shall be deemed a tun of wine, and that 126 such gallons shall be deemed a butt or pipe of wine, and that 63 such gallons shall be deemed an hogshead of wine.

By an act of 13 William III. ch. 5, in 1701, the Winchester bushel had been declared the standard for the measure of grain; and any cylindrical vessel of $18\frac{1}{2}$ inches diameter and 8 inches deep, was made a legal bushel. By a subsequent statute of 12 Anne, ch. 17 sec. 11, the bushel for measuring coal was to be of $19\frac{1}{2}$ inches diameter from outside to outside, and was to contain a quart of water more than the Winchester bushel; which made it of 2217.62 cubical inches.

There are several late acts of parliament (1805, 45 George III.) which mention $272\frac{1}{4}$ cubic inches as the contents of the Winchester gallon, making a bushel of 2178 inches; and others which recognize the existence of measures different from any of the legal standards of the exchequer. By an act of 31 Geo. III. ch. 3, inspectors of corn returns are to make a comparison between the Winchester bushel and the measure commonly used in the city or town of their inspection, and to cause a statement in writing of such comparison to be hung up in some conspicuous place.

By these successive statutes, determining in cubic inches the capacity of the vessels by which certain specific articles shall be measured, the measures bearing the same denomination, but of different contents, are multiplied; and every remnant of the original uniformity of proportion has disappeared, with the exception of that between the wine and ale gallons, and that between the troy and avoirdupois weights.

By the English system of weights and measures before the statute of 1496, the London quarter of a ton was the one measure, to which the bushel for corn, the gallon, deduced by measure, for ale, and the gallon, deduced by weight, for wine, were all referred. The hogshead was a vessel deduced from the cubing of linear measure, containing sixty-three gallons, and measuring eight cubic feet. The gallon thus formed, contained 219.43 cubic inches. This wine gallon, by another law, was to contain eight twelve ounce pounds of wheat. One such pound of wheat, therefore, occupied 27.45 cubic inches. The vessel of eight times 27.45 cubic inches filled with wine, the liquor would weigh 54,857.1 grains of troy weight: and the weight of eight such gallons of wine would be 438,856.8 grains troy. The specific gravity of wine being to that of wheat as 175 to 143, the bushel thus formed would be of 2148.5 cubic inches; and its eighth part, or ale gallon, would be 268.5 inches. This is only two inches more than the standard Winchester bushel of the exchequer was found to contain, and two inches less than the bushel as prescribed by the act of 13 William III.; a difference which a variation in the temperature of the atmosphere is of itself adequate to produce. It proves, that the Winchester bushel has not without reason been preserved as the favorite of all standards, in spite of all the changes, errors, and inconsistencies, of legislation. But it also proves, that the ale and corn gallon ought to have continued as they originally were, of $268\frac{1}{2}$ inches, and the wine gallon of $219\frac{1}{2}$.

The troy and avoirdupois weights are in the proportions to each other of the specific gravity of wheat and of spring water. The twelve and fifteen ounce easterling pounds were intended to be proportional between the gravity of wheat and wine. But they were roughly settled proportions, estimating the weight of wheat to be to that of wine as four to five, and the gravity of wine and of water to be the same. Under the statute of 1496, the wine gallon was of 224 inches. If troy weight was to be introduced, a gallon of this capacity had the great advantage upon which the proportion of uniformity had originally been established. The gallon contained exactly eight pounds avoirdupois of wine. The pint of wine, was a pound of wine. The corn gallon of 272 inches, corresponding with it, had the same advantage. It was filled with eight pounds of corn: a pint of wheat, was a pound of wheat; and the bushel of 2176 inches contained 64 pounds avoirdupois of that wheat, 32 kernels of which weighed one pennyweight troy. But the hogshead, being of eight cubic feet, could have contained only 61¾ gallons, and the ton would have been of 247.

The wine and ale gallons, now established by law, of 231 and 282 inches, are still in the same proportion to each other as the troy and avoirdupois weights: but neither of them is in any useful proportion to the bushel. The corn gallon only is in proportion to the bushel. Neither the wine nor the corn gallons are in any useful proportion either to the weights or the coins. But the troy and avoirdupois weights are, with all the exactness that can be desired, standards for each other: and the cubic foot of spring water weighs exactly 1000 ounces avoirdupois, by which means the ton, of thirty-two cubic feet measure, is in weight exactly 2000 pounds avoirdupois.

Such was originally the system of English weights and measures, and such is it now in its ruins. The substitution of cubic inches, to settle the dimensions of the gallons and bushels, which began with the last century, was a change of the *test* of their contents from gravity to extension. They had before been measured by number, weight, and measure: they are now measured by measure alone. This change has been of little use in promoting the principle of uniformity. As it respects the natural standard, it has only been a change from the weight of a kernel of wheat to the length of a kernel of barley: and although it has specified the particular standard bushels and gallons, selected among the variety, which the inconsistencies of former legislation had produced, it has very unnecessarily brought in a third gallon measure quite incompatible with the primitive system; and it has legalized two bushels of different capacity, so slightly different as to afford every facility to the fraudulent substitution of the one for the other; yet, in the measurement of quantities, resulting in a difference of between three and four per cent.

No further change in this portion of English legislation has yet been made. But the philosophers and legislators of Britain have never ceased to be occupied upon weights and measures, nor to be stimulated by the passion for uniformity. In speculating upon the theory, and in making experiments upon the existing standards of their weights and measures, they seem to have considered the principle of *uniformity* as exclusively applicable to identity, and to have overlooked or disregarded the uniformity of proportion. They found a great variety of standards differing from each other: and instead of searching for the causes of these varieties in the errors and mutability of the law, they ascribed them to the want of an immutable standard from *nature*. They felt the convenience and the facility of decimal arithmetic for *calculation;* and they thought it susceptible of equal application to the divisions and multiplications of *time, space,* and *matter.* They despised the primitive standards assumed from the stature and proportions of the human body. They rejected the secondary standards, taken from the productions of nature most essential to the subsistence of man; the articles for ascertaining the quantities of which, weights and measures were first found necessary. They tasked their ingenuity and their learning to find, in matter or in motion, some *immutable* standard of linear measure, which might be assumed as the single universal standard from which all measures and all weights might be derived. In the review of the proceedings in France relative to this subject, we shall trace the progress and note the results hitherto of these opinions, which have there been embodied into a great and beautiful system. In England they have been indulged with more caution, and more regard to the preservation of existing things.

From the year 1757 to 1764, in the years 1789 and 1790, and from the year 1814 to the present time, the British parliament have, at three successive periods, instituted inquiries into the condition of their own weights and measures, with a view to the reformation of the system, and to the introduction and establishment of greater uniformity. These inquiries have been pursued with ardor and perseverance, assisted by the skill of their most eminent artists, by the learning of their most distinguished philosophers, and by the cotemporaneous admirable exertions, in the same cause of uniformity, of their neighbouring and rival nation.

Nor have the people, or the Congress of the United States, been regardless of the subject, since our separation from the British empire. In their first confederation, these associated states, and in their present national constitution, the people, that is, on the only two occasions upon which the collective voice of this whole Union, in its constituent character, has spoken, the power of *fixing* the standard of weights and measures throughout the United States has been committed to Congress. A report, worthy of the illustrious citizen by whom it was prepared, and, embracing the principles most essential to uniformity, was presented in obedience to a call from the House of Representatives of the first Congress of the United States. The eminent person who last presided over the Union, in the parting message by which he announced his intention of retiring from public life, recalled the subject to the attention of Congress with a renewed recommendation to the principle of decimal divisions. Elaborate reports, one from a committee of the Senate in 1793, and another from a committee of the House of Representatives, at a recent period, have since contributed to shed further light upon the subject: and the call of both Houses, to which this report is the tardy, and yet too early answer, has manifested a solicitude for the improvement of the existing system, equally earnest and persevering with that of the British parliament, though not marked with the bold and magnificent characters of the concurrent labors of France.

After a succession of more than sixty years of inquiries and experiments, the British parliament have not yet acted in the form of law. After nearly forty of the same years of separate pursuit of the same object, *uniformity,* the Congress of the United States have shown the same cautious deliberation: they have yet authorized no change of the existing law. That neither country has yet changed its law, is, perhaps, a fortunate circumstance, in reference to the principle of uniformity, for both. If this report were authorized to speak to both nations, as it is required to speak to the legislature of one of them, on a subject in which the object of pursuit is the same for both, and the interest in it common to both, it would say—Is your object *uniformity?* Then, before you change any part of your system, such as it is, compare the uniformity that you must lose, with the uniformity that you may gain, by the alteration. At this hour, fifteen millions of Britons, who, in the next generation, may be twenty, and ten millions of Americans, who, in less time, will be as many, have the same legal system of weights and measures. Their mile, acre, yard, foot, and inch—their bushel of wheat, their gallon of beer, and their gallon of wine, their pound avoirdupois, and their pound troy, their cord of wood, and their ton of shipping, are the same. They are of the nations of the earth, the two, who have with each other the most of that intercourse which requires the constant use of weights and measures. Any change whatever in the system of the one, which would not be adopted by the other, would destroy all this existing uniformity. Precious, indeed, must be that uniformity, the mere promise of which, obtained by an alteration of the law, would more than compensate for the abandonment of this.

If these ideas should be deemed too cold and cheerless for the spirit of theoretical improvement; if Congress should deem their powers competent, and their duties imperative, to establish uniformity as respects weights and measures in its most universal and comprehensive sense; another system is already made to their hands. If that universal uniformity, so desirable to human contemplation, be an obtainable perfection, it is now attainable *only* by the adoption of the new French system of metrology, in all its important parts. Were it even possible to construct another system, on different principles, but embracing in equal degree all the great elements of uniformity, it would still be a system of diversity with regard to France, and all the followers of her system. And as she could not be expected to abandon that, which she has established at so much expense, and with so much difficulty, for another, possessing, if equal, not greater advantages, there would still be two rival systems, with more desperate chances for the triumph of uniformity by the recurrence to the same standard of all mankind.

The system of modern France originated with her Revolution. It is one of those attempts to improve the condition of human kind, which, should it even be destined ultimately to fail, would, in its failure, deserve little less admiration than in its success. It is founded upon the following principles:

1. That all weights and measures should be reduced to one *uniform* standard of linear measure.
2. That this standard should be an aliquot part of the circumference of the globe.
3. That the unit of linear measure, applied to matter, in its three modes of extension, length, breadth, and thickness, should be the standard of all measures of length, surface, and solidity.

4. That the cubic contents of the linear measure, in distilled water, at the temperature of its greatest contraction, should furnish at once the standard weight and measure of capacity.
5. That for every thing susceptible of being measured or weighed, there should be only one measure of length, one weight, one measure of contents, with their multiples and subdivisions exclusively in decimal proportions.
6. That the principle of decimal division, and a proportion to the linear standard, should be annexed to the coins of gold, silver, and copper, to the moneys of account, to the division of *time,* to the barometer and thermometer, to the plummet and log lines of the sea, to the geography of the earth and the astronomy of the skies; and, finally, to every thing in human existence susceptible of comparative estimation by weight or measure.
7. That the whole system should be equally suitable to the use of all mankind.
8. That every weight and every measure should be designated by an appropriate, significant, characteristic name, applied exclusively to itself.

This system approaches to the ideal perfection of *uniformity* applied to weights and measures; and, whether destined to succeed, or doomed to fail, will shed unfading glory upon the age in which it was conceived, and upon the nation by which its execution was attempted, and has been in part achieved. In the progress of its establishment there, it has been often brought in conflict with the laws of physical and of moral nature; with the impenetrability of matter, and with the habits, passions, prejudices, and necessities, of man. It has undergone various important modifications. It must undoubtedly still submit to others, before it can look for universal adoption. But, if man upon earth be an improveable being; if that universal peace, which was the object of a Saviour's mission, which is the desire of the philosopher, the longing of the philanthropist, the trembling hope of the Christian, is a blessing to which the futurity of mortal man has a claim of more than mortal promise; if the Spirit of Evil is, before the final consummation of things, to be cast down from his dominion over men, and bound in the chains of a thousand years, the foretaste here of man's eternal felicity; then this system of common instruments, to accomplish all the changes of social and friendly commerce, will furnish the links of sympathy between the inhabitants of the most distant regions; the metre will surround the globe in use as well as in multiplied extension; and one language of weights and measures will be spoken from the equator to the poles.

The establishment of this system of metrology forms an era, not only in the history of weights and measures, but in that of human science. Every step of its progress is interesting: and as a statement of all the regulations in France concerning it is strictly within the scope of the requisitions of both Houses, a rapid review of its origin, progress, and present state, with due notice of the obstacles which it has encountered, the changes through which it has passed, and its present condition, is deemed necessary to the performance of the duty required by the call.

In the year 1790, the present prince de Talleyrand, then bishop of Autun, distributed among the members of the constituent assembly of France a proposal, founded upon the excessive diversity and confusion of the weights and measures then prevailing all over that country, for the reformation of the system, or rather, for the foundation of a new one upon the principle of a single and universal standard. After referring to the two objects which had previously been suggested by Huyghens and Picard, the pendulum and the proportional part of the circumference of the earth, he concluded by giving the preference to the former, and presented the project of a decree. First, that exact copies of all the different weights and elementary measures, *used* in every town of France, should be obtained and sent to Paris: Secondly, that the national assembly should write a letter to the British parliament, requesting their concurrence with France in the adoption of a natural standard for weights and measures, for which purpose Commissioners, in equal numbers from the French Academy of Sciences and the British Royal Society, chosen by those learned bodies, respectively, should meet at the most suitable place, and ascertain the length of the pendulum at the 45th degree of latitude, and from it an invariable standard for all measures and weights: Thirdly, that, after the accomplishment, with all due solemnity, of this operation, the French Academy of Sciences should fix with precision the tables of proportion between the new standards and the weights and measures previously used in the various parts of France; and that every town should be supplied with exact copies of the new standards, and with tables of comparison between them and those of which they were to supply the place. This decree, somewhat modified, was adopted by the assembly, and, on the 22d of August, 1790, sanctioned by Louis the Sixteenth. Instead of writing to the British parliament themselves, the assembly requested the king to write to the king of Great Britain, inviting him to propose to the parliament the formation of a joint commission of members of the Royal Society and of the Academy of Sciences, to ascertain the natural standard in the length of the pendulum. Whether the forms of the British constitution, the temper of political animosity then subsisting between the two countries, or the convulsions and wars which soon afterwards ensued, prevented the acceptance and execution of this proposal, it is deeply to be lamented that it was not carried into effect. Had the example once been set of a *concerted* pursuit of the great common object of *uniformity* of weights and measures, by two of the mightiest and most enlightened nations upon earth, the prospects of ultimate success would have been greatly multiplied. By no other means can the uniformity, with reference to the persons using the same system, be expected to prevail beyond the limits of each separate nation. Perhaps when the spirit which urges to the improvement of the social condition of man, shall have made further progress against the passions with which it is bound, and by which it is tramelled, then may be the time for reviving and extending that generous and truly benevolent proposal of the constituent national assembly of France, and to call for a *concert* of civilized nations to establish one uniform system of weights and measures for them all.

The idea of associating the interests and the learning of other nations in this great effort for common improvement was not confined to the proposal for obtaining the concurrent agency of Great Britain. Spain, Italy, the Netherlands, Denmark, and Switzerland, were actually represented in the proceedings of the Academy of Sciences to accomplish the purposes of the national assembly. But, in the first instance, a committee of the Academy of Sciences, consisting of five of the ablest members of the academy and most eminent mathematicians of Europe, Borda, Lagrange, Laplace, Monge, and Condorcet, were chosen, under the decree of the assembly, to report to that body upon the selection of the natural standard, and other principles proper for the accomplishment of the object. Their report to the academy was made on the 19th of March, 1791, and immediately transmitted to the national assembly, by whose orders it was printed. The committee, after examining three projects of a natural standard, the pendulum beating seconds, a quarter of the equator, and a quarter of the meridian, had, on full deliberation, and with great accuracy of judgment, preferred the last; and proposed, that its ten millionth part should be taken as the standard unit of linear measure; that, as a second standard of comparison with it, the pendulum vibrating seconds at the 45th degree of latitude should be assumed; and that the weight of distilled water at the point of freezing, measured by a cubical vessel in decimal proportion to the linear standard, should determine the standard of weights and of vessels of capacity.

For the execution of this plan, they proposed six distinct scientific operations, to be performed by as many separate committees of the academy.

1. To measure an arc of the meridian from Dunkirk to Barcelona, being between nine and ten degrees of latitude, including the 45th, with about six to the north and three to the south of it, and to make upon this line all suitable astronomical observations.

2. To measure anew the bases which had served before for the admeasurement of a degree in the construction of the map of France.

3. To verify, by new observations, the series of triangles, which had been used on the former occasion, and to continue them to Barcelona.

4. To make, at the 45th degree of latitude, at the level of the sea, in vacuo, at the temperature of melting ice, observations to ascertain the number of vibrations in a day of a pendulum equal to the ten millionth part of the arc of the meridian.

5. To ascertain, by new experiments, carefully made, the weight in vacuo of a given mass of distilled water at the freezing point.

6. To form a scale and tables of equalization between the new measures and weights proposed, and those which had been in common use before.

This report was sanctioned by a decree of the assembly: and four committees of the academy were appointed; the three first of those enumerated objects having been intrusted to one committee, consisting of Mechain and Delambre. The experiments upon the pendulum were committed to Borda, Mechain, and Cassini; those on the weight of water to Lefevre Gineau, and Fabbroni; and the scale and tables to a large committee on weights and measures.

The performance of all these operations was the work of more than seven years. Two of them, the measurement of the arc of the meridian, and the ascertainment of the specific gravity of water in vacuo, were works requiring that combination of profound learning which is possessed of the facts in the recondite history of nature already ascertained, with that keenness of observation which detects facts still deeper hidden; that fertility of genius which suggests new expedients of invention, and that accuracy of judgment, which turns to the account, not only of the object immediately sought, but of the general interests of science, every new fact observed. The actual admeasurement of an arc of the meridian of that extent had never before been attempted. The weighing of distilled water in vacuo had never before been effected with equal accuracy. And, in the execution of each of these works, nature, as if grateful to those exalted spirits who were devoting the labors of their lives to the knowledge of her laws, not only yielded to them the object which they sought, but disclosed to each of them another of her secrets. She had already communicated by her own inspiration to the mind of Newton, that the earth was not a perfect sphere, but an oblate spheroid, flattened at the poles: and she had authenticated this discovery by the result of previous admeasurements of degrees of the meridian in different parts of the two hemispheres. But the proportions of this flattening, or, in other words, the difference between the circles of the meridian and equator, and between their respective diameters, had been variously conjectured, from facts previously known. To ascertain it with greater accuracy was one of the tasks assigned to Delambre and Mechain; for, as it affected the definite extension of the meridian circle, the length of the metre, or aliquot part of that circle which was to be the standard unit of weights and measures, was also proportionably affected by it. The result of the new admeasurement was, to show that the flattening was of $\frac{1}{334}$; or that the axis of the earth was to the diameter of the equator as 333 to 334. Is this proportion to the decimal number of 1,000 accidental? It is confirmed as matter of fact, by the existing theories of astronomical nutation and precession, as well as by experimental results of the length of the pendulum in various latitudes. Is it also an index to another combination of extension, specific gravity, and numbers, hitherto undiscovered? However this may be, the fact of the proportion was, on this occasion, the only object sought. This fact was attested by the diminution of each degree of latitude, in the movement from the north to the equator; but the same testimony revealed the new and unexpected fact, that the diminution was not regular and gradual, but very considerably different at different stages of the progress in the same direction; from which the inference seems conclusive, that the earth is no more in its breadth than in its length, perfectly spherical, and that the northern and southern hemispheres are not of dimensions precisely equal.

The other discovery was not less remarkable. The object to be ascertained was the specific gravity of a given mass of water in vacuo, and at its *maximum* of density; that is, at the temperature where it weighs most in the smallest space. That fluids are subject to the general laws of expansion and contraction from heat and cold, was the principle upon which the experiments were commenced. It was also known that, in the transition of fluids to a solid state, the reverse of this phenomenon occurs, and that water, in turning to ice, instead of contracting, expands. It had been supposed that the freezing point was that at which this polarity of heat and cold, if it may be so called, was inverted, and that water, contracting as it cooled until then, began at once to freeze and to expand. The discovery made by Lefevre Gineau, and Fabbroni, was, that the change took place at an earlier period; that water contracts as it cools, till at five degrees above 0 of the centigrade, answering to forty-one of Fahrenheit's thermometer, and, from that term, gradually expands as it grows cold, till fixed in ice at 0 of the former, or thirty-two degrees of the latter.

In the admeasurement of the arc of the meridian, and in the weighing of the given volume of water, the standard measure and weights, previously established by the laws of France, were necessarily used. The identical measure was a toise or fathom belonging to the Academy of Sciences, which had been used for the admeasurement of several degrees of the meridian between the years 1737, and 1741, in Peru, and had thence acquired the denomination of the *Toise du Perou.* In 1766 it had served as the standard from which eighty others had been copied, and sent to the principal bailiwicks in France, and to the chatelet at Paris. The instruments used by Delambre and Mechain, for their mensurations, were two platina rods, each of double the length of this fathom of Peru. A repeating circle, a levelling instrument, and a metallic thermometer, consisting of two blades, one of brass, and the other of platina, and calculated to show the difference of expansion produced upon the two metals by the ordinary alternations of heat and cold in the atmosphere, all invented by the ingenious and skilful artist Borda, were also among the instruments used by the commissioners.

The weights with which the new standard was compared, were a pile of fifty marks, or twenty-five Paris pounds, called the weights of Charlemagne, and which, though not of the antiquity of that prince's age, had been used as standards for a period of more than five hundred years.

The fathom of Peru was divided into six standard royal feet of France, each foot into twelve thumbs, each thumb into twelve lines. The toise, therefore, was of seventy-two thumbs, or 864 lines. The standard metre of platina, the ten millionth part of the quarter of the meridian, measured by the brass fathom of Peru, was found to be equal to 443 lines, and 295,936 decimal parts of a line: and as it was found impossible to fix in the concrete form a division smaller than the thousandth part of a line, the definitive length of the metre was fixed at 443,296 lines, equivalent, by subsequent experiments of the academy, to 39.3827 English inches; by the latest experiments of captain Kater to 39.37079; and by those of Mr. Hassler, in this country, to 39.3802.

The Paris pound, mark-weight as it was called, (poids de marc,) of the pile of Charlemagne, consisted of two marks, each mark of eight ounces, each ounce of eight gros or drams, each gros of three deniers or pennyweights, and each denier of twenty-four grains. The pound, therefore, consisted of 9,216 grains, and was equal to fifteen ounces and fifteen penny weights, or 7,560 grains troy. The grain was rather more than four-fifths of the troy grain, and had probably, in the origin, been equivalent to the kernel of wheat, which the troy grain could scarcely have been. The cubic decimetre, or tenth part of the metre, of distilled water, at the temperature of its greatest density, weighed in vacuo, was found of equal weight with 18.827 grains $\frac{15}{100}$ of a grain; or two pounds, five gros, thirty-five grains $\frac{15}{100}$ of the mark weight: and this, by the name of the kilogramme was made the standard weight, its thousandth part being the gramme, or unit, equivalent to 15.44572 grains troy, or about two and one-fifth pounds avoirdupois.

The capacity of the vessel containing this water was at the same time made the standard of all measures, liquid or dry: it was called a litre, and is of the contents of 61.0271 cubic inches, about one-twentieth more than our wine quart. The metre was applied to superficial and solid measures, according to their proportions: the chain of ten metres being applied to land measure, and its square denominated an *arc;* the cubic metre was called a *stere.*

The principle of decimal arithmetic was applied exclusively to all these weights and measures: their multiples were all tenfold, and their subdivisions were all tenth parts.

To complete the system, a vocabulary of new denominations was annexed to every weight and measure belonging to it. As a circumstance of great importance to the final success of the system, it may be remarked that these two incidents, the exclusive adoption of decimal divisions, and the new nomenclature, have proved the greatest obstacles to the general introduction of the new weights and measures among the people.

It has indeed from its origin, like all great undertakings, been obliged to contend with the intemperate zeal and precipitation of its friends, not less than with prejudice, ignorance, and jealousy, of every description. The admeasurement of the meridian was commenced at the very moment of the fanatical paroxysm of the French revolution. At every station of their progress in the field survey, the commissioners were arrested by the suspicions and alarms of the people, who took them for spies, or engineers of the invading enemies of France. The government was soon overthrown; the Academy of Sciences abolished; and the national assembly of the first constitutional monarchy, just at the eve of their dissolution, instead of waiting calmly for the completion of the great work which was to lay the foundation for a system to be as lasting as the globe, in a fit of impatience passed, on the 1st of August, 1793, a law declaring that the system should go immediately into operation, and assuming for the length of the standard metre the ten millionth part of the quadrant of the meridian according to the result of the old admeasurement of a degree in 1740, and arranging an entire system of weights and measures, in decimal divisions, with new denominations, all of which were to be merely temporary, and to cease when the definitive length of the metre should be ascertained. This extraordinary act was probably intended, as it directly tended, to prevent the further prosecution of the original plan: and though, soon after, it was followed by a decree of 11th September, 1793, authorizing the temporary continuance of the general committee of weights and measures, which had been appointed by the academy, yet, on the 23d December of the same year, a decree of Robespierre's committee of public safety dismissed from the commission Borda, Lavoisier, Laplace, Coulomb, Brisson, and Delambre, on the pretence that they were not republicans sufficiently pure. Mechain escaped the same proscription only because he was detained as a prisoner in Spain.

Yet even Robespierre and his committee were ambitious, not only of establishing the system of new weights and measures in France, but of offering them to the adoption of other nations. By a decree of that committee of 11th December, 1793, the board or commission of weights and measures were directed to send to the United States of America a metre in copper and a weight, being copies of the standards then just adopted. They were accordingly transmitted: and on the 2d of August, 1794, the two standards were, by the then French minister plenipotentiary Fauchet, sent to the Secretary of State, with a letter, recommending, with some urgency, the adoption of the system by the United States. This letter was communicated to Congress by a message from the President of the United States, of the 8th of January, 1795.

In the mean time the mensuration of the arc of the meridian was entirely suspended by the dismissal of Delambre, and the detention of Mechain. Its progress was renewed by a decree of the national convention of 7th April, 1795, (18 Germinal, An. 3) which abolished almost entirely the nomenclature of the temporary standards adopted in August, 1793, and substituted a new one, being that still recognized by the law, and the units of which have been already mentioned; the metre, the gramme, the arè, the litre, and the stere. To express the multiples of these units, the Greek words denominating ten, a hundred, a thousand, and ten thousand, were prefixed as additional syllables, while their tenth, hundredth, and thousandth parts were denoted by similar prefixed syllables from the Latin language. Thus, the myria-metre is ten thousand, and the kilo-metre one thousand, the hecto-metre one hundred, and the deca-metre ten metres; each of those prefixed syllables being the Greek word expressive of those respective numbers ; while the deci-metre, the centi-metre, and the milli-metre, are tenth, hundredth, and thousandth parts, signified by the Latin syllables respectively prefixed to them. The theory of this nomenclature is perfectly simple and beautiful. Twelve new words, five of which denote the things, and seven the numbers, include the whole system of metrology; give distinct and significant names to every weight, measure, multiple, and subdivision, of the whole system ; discard the worst of all the sources of error and confusion in weights and measures, the application of the same name to different things ; and keep constantly present to the mind the principle of decimal arithmetic, which combines all the weights and measures, the proportion of each weight or measure with all its multiples and divisions, and the chain of uniformity which connects together the profoundest researches of science with the most accomplished labors of art and the daily occupations and wants of domestic life in all classes and conditions of society. Yet this is the part of the system which has encountered the most insuperable obstacles in France. The French nation have refused to learn, or to repeat these twelve words. They have been willing to take a total and radical change of things; but they insist upon calling them by old names. They take the metre; but they must call one-third part of it a foot. They accept the kilogramme ; but, instead of pronouncing its name, they choose to call one half of it a pound. Not that the third of a metre is a foot, or the half of a kilogramme is a pound; but because they are not very different from them, and because, in expressions of popular origin, distinctness of idea in the use of language is more closely connected with habitual usage than with precision of expression.

This observation may be illustrated by our own experience, in a change effected by ourselves in the denominations of our coins, a revolution by all experience known to be infinitely more easy to accomplish than that of weights and measures. At the close of our war for independence, we found ourselves with four English words, pound, shilling, penny, and farthing, to signify all our moneys of account. But, though English words, they were not English things. They were no where sterling: and scarcely in any two states of the Union were they representatives of the same sums. It was a Babel of confusion by the use of four words. In our new system of coinage we set them aside. We took the Spanish piece of eight, which had always been the coin most current among us, and to which we had given a name of our own—a dollar. Introducing the principle of decimal divisions, we said, a tenth part of our dollar shall be called a *dime*, a hundredth part a *cent*, and a thousandth part a *mille*. Like the French, we took all these new denominations from the Latin language; but instead of prefixing them as syllables to the generic term dollar, we reduced them to monosyllables, and made each of them significant by itself, without reference to the unit of which they were fractional parts. The French themselves, in the application of their system to their coins, have followed our example; and, assuming the franc for their unit, call its tenth part a *decime*, and its hundredth a *centime*. It is now nearly thirty years since our new moneys of account, our coins, and our mint, have been established. The dollar, under its new stamp, has preserved its name and circulation. The cent has become tolerably familiarized to the tongue, wherever it has been made by circulation familiar to the hand. But the dime having been seldom, and the mille never, presented in their material images to the people, have remained so utterly unknown, that now, when the recent coinage of dimes is alluded to in our public journals; if their name is mentioned, it is always with an explanatory definition to inform the reader, that they are ten cent pieces; and some of them which have found their way over the mountains, by the generous hospitality of the country, have been received for more than they were worth, and have passed for an eighth, instead of a tenth, part of a dollar. Even now, at the end of thirty years, ask a tradesman, or shopkeeper, in any of our cities what is a dime or a mille, and the chances are four in five that he will not understand your question. But go to New York and offer in payment the Spanish coin, the unit of the Spanish piece of eight, and the shop or market-man will take it for a *shilling*. Carry it to Boston or Richmond, and you shall be told it is not a shilling, but nine pence. Bring it to Philadelphia, Baltimore, or the City of Washington, and you shall find it recognized for an eleven-penny bit; and if you ask how that can be, you shall learn that, the dollar being of ninety pence, the eighth part of it is nearer to eleven than to any other number: and pursuing still further the arithmetic of popular denominations, you will find that half eleven is five, or, at least, that half the eleven-penny bit is the fi-penny bit, which fi-penny bit at Richmond shrinks to four pence half-penny, and at New York swells to six pence. And thus we have English denominations most absurdly and diversely applied to Spanish coins; while our own lawfully established dime and mille remain, to the great mass of the people, among the hidden mysteries of political economy—state secrets.

Human nature, in its broadest features, is every where the same. This result of our own experience, upon a small scale, and upon a single object, will easily account for the repugnance of the French people to adopt the new nomenclature of their weights and measures. It is not the length of the words that constitutes the objection against them, nor the difficulty of pronunciation; for, fi-penny bit is as hard to speak and as long a word as kilogramme, and eleven-penny bit has certainly more letters and syllables, and less euphony, than myria-metre. But it is because, in the ordinary operations of the mind, distinctness of idea is, by the laws of nature, linked with the chain of association between sensible images and their habitual denominations, more closely than with the exactness of logical analysis.

The nomenclature of the French metrology was established by the law of 7th April, 1795, although the metre and the kilogramme were only provisional and not definitive. It was known that the difference between the provisional and definitive metre and kilogramme would be very small, scarcely perceptible: and by that inverted logic which presides over all precipitate legislation, it was concluded that because it was small it would be unimportant: instead of which, sound reason would have inferred, that, to the purpose of uniformity, the smaller the difference was, the greater was the danger of its producing confusion between the temporary and the perpetual things which were to bear the same name.

But with the hasty call for a provisional metre and kilogramme, the law of 7th April, 1795, gave the definitive nomenclature, and directed the renewal of all the operations commenced under the direction of the Academy of Sciences; and the persons employed upon them were reinstated in their functions, by the committee of public instruction of the national convention. A commission of twelve persons, Berthollet, Borda, Brisson, Coulomb, Delambre, Hauy, Lagrange, Laplace, Mechain, Monge, Prony, and Vandermonde, was appointed on the 17th of April, 1795, for the final accomplishment of the original plan; the most important and laborious part of which, the admeasurement of the arc of the meridian, was immediately resumed by Delambre and Mechain. By them the whole distance from Dunkirk to Mont Jouy, near Barcelona, a distance of nine degrees and two thirds, more than a tenth part of the quadrant of the meridian, was measured by trigonometrical survey. The angles formed by every station with those next before and after it, rectified by the angles of elevation and depression formed by the inequalities on the surface of the ground, to reduce the whole to the level of the horizon, were measured and referred to the measure of two bases, one between Melun and Lieusaint, the other between Vernet and Salces, near Perpignan; each serving as a corrective upon the other. Observations of azimuth ascertained the direction of the sides of the triangles, with reference to the meridian; and astronomical observations ascertained the celestial arc, corresponding with that which was measured upon the earth. The distance from Dunkirk to Rhodez, about 450 miles, was performed by Delambre; and that from Barcelona to Rhodez, upwards of 200 miles, by Mechain. The base of Melun was of 6075.90 and that of Perpignan 6006.25 toises, each nearly seven miles: and though at the distance of near 400 miles from each other, the base of Perpignan, calculated by inference from the chain of triangles between them, differed from its actual admeasurement less than one foot. The portion of the distance allotted to Mechain was less than one third of the whole; but, traversing the Pyrenees, and being chiefly upon the Spanish territories, was attended with more difficulties than those encountered by his associate. Mechain, in the execution of his task, had formed the project of extending the survey to the Balearic islands, which would have made the portion of the arc south of the forty-fifth degree equal to that northward of it. With a firmness and perseverance of pursuit, amidst innumerable obstacles, he had proceeded far in the execution of this supplementary plan. His triangles were already extended from Barcelona to Tortosa. His stations had been selected to Cullora. Six or seven triangles more would have carried his work to its termination in the island of Ivica. Arrested by a fever, in his first progress, and compelled then to abandon the attempt, he had resumed it after the result of the original plan had been ascertained, and the new system had been finally established by law. The hand which sets bounds to all human pursuits again and definitively met and closed his career. He died on the 20th of September, 1805, at Castellon de la Plana, in the Spanish province of Valencia. His more fortunate associate, Delambre, has published, in three quarto volumes, under the title of the " Basis of the metrical decimal system, or measure of the arc of " the meridian between Dunkirk and Barcelona," all the details and results of this admirable operation. A fourth volume yet remains to be published, which will contain the account of the actual execution since the death of Mechain, of the idea which he had conceived of extending the admeasurement to the island of Formentara, and of the additional extension of it northward to the Shetland islands, by connecting it with the trigonometrical survey of Great Britain. This work, in passing to future ages, a monument of the philosophy, science, public spirit, and active benevolence of our own, will redeem, by the martyrdom of genius and learning in the cause of human happiness, the blood-polluted glories of cotemporaneous war.

The reports of the proceedings of Delambre and Mechain, as well of their field surveys as of their astronomical observations, and all their calculations, were submitted to the inspection, scrutiny, and revision, of a committee of the mathematical and physical class of the national institute, that phœnix of science which had arisen from the ashes of the academy of sciences. The observations to ascertain the length of the pendulum. and the experiments for determining the specific gravity of distilled water at its maximum of density, were submitted to the same ordeal. Two reports upon the whole result were made to the class, one by Trallés of the Helvetic Confederation, the other by Van Swinden of the Netherlands, two of the foreign associates, who had been invited to co-operate in the labor, and to participate in the honor of the undertaking. These two reports, combined by Van Swinden into one, were then reported from the class to the general meeting of the institute, and by that body, with all suitable solemnity, to the two branches of the national assembly of France, on the 22d of June, 1799, together with a definitive metre of platina, made by Lenoir, and a kilogramme of the same metal, made by Fortin. They were introduced by an appropriate address at the bar of the two houses, by the presiding member of the national institute, La Place; to which answers were returned by the respective presidents of the two legislative chambers. On the same day, the standard metre and kilogramme were deposited in the hands of the keeper of the public archives; and a record of the fact was made and signed by him, and by all the members of the institute, foreign associates, and artists, whose joint labors had contributed to the consummation of this more than national undertaking.

No apology will be deemed necessary by Congress for dwelling upon these details which signalized the establishment of the new French metrical system. The spectacle is at once so rare and so sublime, in which the genius, the science, the skill, and the power of great confederated nations are seen joining hand in hand in the true spirit of fraternal equality, arriving in concert at one destined stage of improvement in the condition of human kind; that, not to pause for a moment, were it even from occupations not essentially connected with it, to enjoy the contemplation of a scene so honorable to the character and capacities of our species, would argue a want of sensibility to appreciate its worth. This scene formed an epocha in the history of man. It was an example and an admonition to the legislators of every nation, and of all after-times.

On the 10th of December, 1799, (19 Frimaire 8,) the temporary metre and kilogramme, which had been ordained by the laws of 1st August, 1793, and 7th April, 1795 (18 Germinal 3,) were abolished. That metre had been of 443 lines and $\frac{44}{100}$ of a line of the ancient foot, standarded by the fathom of Peru. The new and definitive metre was fixed at 443 lines $\frac{296}{1000}$. The difference between them was about $\frac{1}{7}$ of a line, or $\frac{1}{100}$ of our inch, a difference imperceptible for all ordinary uses; but very important as a standard variety, and immediately apparent when multiplied to the cube for the measure of capacity and the weight. Thus the temporary kilogramme had been of 18,841 grains mark weight, while the new and definitive kilogramme was reduced to 18,827.15 grains.

During the violent ebullitions of the most inflammatory period of the French Revolution, it had been imagined that in the reformation of the system of weights and measures, upon the principle of uniformity, the mensuration of time ought to be included. But this was a different project from that of the reformed metrology, originating in motives less pure and ingenuous, connected with purposes interfering with religious impressions, and quite inconsistent with one of the principal expedients of perpetuating the identity of the new weights and measures. The length of the pendulum beating seconds, it has been seen, is, in the new metrical system, the test of verification for that of the metre, in case the original platina standard should be lost. The pendulum beating seconds vibrates 86,400 times in the solar day of 24 hours. But, the fiery spirits of the Revolution called for a reformation of the calendar, for a new constitution of the seasons, and above all, for decimal divisions. The establishment of the French republic was a new æra to the world. It had taken place on the 22d of September, 1792, the day of the autumnal equinox, when the sun entered the sign of the Balance, the symbol of equality. Before it the Christian æra was to disappear. The new year was to commence with that day. The division of twelve months was to be retained; but they were all to be of three times ten or thirty days. The division of weeks of seven days, beginning with one specially devoted to the worship of the Creator, and repose, was to be abolished; but every tenth day was to be dedicated to some moral abstraction or virtue, such as liberty, equality, fraternity, patriotism, conjugal affection, filial piety, old age, and once a year to the Supreme Creator, whose existence was formally authenticated by a decree of the national convention. After their thirty-six decads, there remained five, and in leap-year six complementary days, to which they gave a name which can scarcely be repeated with decency; but which were to be all holidays, and in which were to be revived the Olympic games of ancient Greece. The names of the months were to be significant. The three successive months, composing each of the four seasons, were to have the same terminating syllable, which in its sound should convey to the ear its distinctive character. The first of the four was *aire*, which was supposed to indicate the solemn and majestic tranquillity of autumn; the second *ose*, a dull and heavy sound, marking the torpor and frigidity of winter; the third *al*, which had all the reviving influence and liquid harmony of spring; and the fourth *dor*, burning to the fancy with the vivid ardors of summer. To these terminating syllables, each month had an appropriate prefix. Thus, in autumn, Vendemi-aire, was the month of vintage; Brum-aire, the month of fogs; and Frim-aire, the month of incipient cold. From the winter solstice to the vernal equinox, there was Niv-ose, the month of snow; Pluvi-ose, the month of rain; and Vent-ose, the month of wind. These were succeeded by the darlings of the year; Germin-al, the month of buds; Flore-al, the month of blossoms; and Prairi-al, the month of blooming meads. The procession closed with the bounties and fervors of summer; Messi-dor, the month of harvests; Thermi-dor, the month of heat; and Fructi-dor, the month of fruit. The days of their decad were to be denominated by their numbers from one to ten; as Primedi, first day, Duodi, second day, and so on to Decadi, the tenth day; which was to be the day of relaxation from labor, and of meditations upon virtue. But the clashing of the new calendar with the new metrology was the division of the solar day, not into 24 hours, of 60 minutes, with 60 seconds to the minute; but into ten hours, each of 100 minutes, and each minute of 100 seconds. The pendulum of that day, therefore, would vibrate 100,000 times, and would be of quite a different length from that which was to be the test of verification to the metre.

This system has passed away, and is forgotten. This incongruous composition of profound learning and superficial frivolity, of irreligion and morality, of delicate imagination and coarse vulgarity, is dissolved. This statue, with the form of Apollo and the face of Silenus, has crumbled into dust; but it was established by a law of the 5th of October, 1792, and for the space of twelve years it was the calendar of the French nation. Henceforth it will only be remembered as preparing future problems in chronology.

The division of the day into a hundred thousand parts had some reasons to recommend it, but was the first part of the system that was abandoned. It had been decreed as compulsory, with the new nomenclature of the calendar, on the 24th of November, 1793, (4 Frimaire 2,) but this regulation was indefinitely suspended by the law of 7th April, 1795, (18 Germinal 3.)

On the 8th of April, 1802, when a First Consul, soon to be for life, had produced some perturbation of that balance, the symbol of equality in which the sun had first shone upon the French Republic, there passed a law, retaining the *equinoctial* or republican calendar for all civil purposes, but resuming the *solstitial* or Gregorian calendar so far as to restore its week of seven days with their names, and its Sabbath of the first of them. The terms *equinoctial* and *solstitial*, in this law, applied to the new and old calendars, seem studiously selected to veil the balance of equality on one side, and the Sabbath of religion on the other.

But on the 9th of September, 1805, (22 Fructidor 13,) in the month of *fruits*, and when the sun of the French Republic had got, if not into the sign, at least deep into the constellation, of the Lion; when the legend of her coins bore on one side the name and head of Napoleon Emperor, and on the other, the name of the French Republic, a senatus consultum ordained, that, from the 11th of that dull and heavy month of snows of the 14th year, the 1st of January, 1806, should re-appear, and the Gregorian calendar should be restored to use throughout the Republican Empire.

The decimal divisions, and the fanciful contexture of the equinoctial calendar, were a sort of episode to the new system of metrology. The attempt to decimate the year in its number of days was equally useless and absurd. The five successive holidays at the close of the year, just at the season of the vintage, with the institution of athletic sports, were a waste of time, and a provocation to mischievous idleness, ill compensated by the retrenchment of sixteen Sundays in the year, at the distance of a week from each other, and devoted to the exercises of piety.

The application of the metrical system to geography and astronomy was a much more rational part of the project; but has been attended with difficulties in execution hitherto insuperable. In adopting an aliquot decimal part of the quadrant of the meridian for the unit of long measures, it formed a natural division of the quadrant itself into ten parts, each of ten degrees. The degree would then have been of 100,000 metres, and the number of degrees to encircle the earth would have been four hundred. The degree, which is now of about sixty-nine English miles, would have been of about sixty-two, and the facility of all astronomical, geographical, and nautical, calculations, would have been much increased. But it would have rendered useless all the tables indispensable to the navigator, astronomer, and geographer; and, if it had not produced the same effect upon all the maps and charts now in use, it would have tended to produce confusion between those of the old and those of the new system. The ancient division of the sphere, and, consequently, of the circle, into 360, and therefore into quadrants of 90 degrees, originated in the coincidence of the daily rotations of the earth in its orbit round the sun, or the apparent motion of the sun in the ecliptic, which, as near as the approximation of numbers can bring it, is of one degree every day. The division of the day into twenty-four hours, each of sixty minutes, is founded on a similar coincidence of time in the rotation of the earth round its axis, and the apparent daily revolution of the firmament round the earth resulting from it; giving for the rising or setting of each sign of the zodiac a term of two hours, and for each degree of the circle described by the earth in its rotation a term of four minutes, or fifteen degrees to the hour. The adoption of the decimal divisions for the quadrant of the meridian, and for the circle, would have disturbed all these harmonies, as well as that of the sexagesimal division of the circle by the radius; a division not perfectly exact, since the radius is not exactly the sixth part of the circumference, but which, having been found the most convenient for practice, has been established from the remotest antiquity, and, being already used by all the civilized nations of the earth, could not, by being set aside, tend to uniformity, unless the method to supply its place could be alike secure of universal adoption.

The divisions of the barometer had always been marked in inches and lines. The application to it of the decimetre, its multiples and divisions, had for observation and calculation the usual conveniences of the decimal arithmetic. The graduation of the thermometer had always been arbitrary and various in different countries. The principle of the instrument was every where the same, that of marking the changes of heat and cold in the atmosphere, by the expansion and contraction which they produced upon mercury or alcohol. The range of temperature between boiling and freezing water was usually taken for the term of graduation, but, by some, it was graduated downwards from heat to cold, and by others upwards from cold to heat. By some the range between the two terminating points was divided into 80, 100, 150, or 212, degrees. One put the freezing, and another the boiling, point at 0. Reaumur's thermometer, used in France, began with 0 for the freezing point, and placed the boiling point at 80. Fahrenheit's, commonly used in England, and in this country, has the freezing point at 32, and the boiling point at 212. The centigrade thermometer, adopted by the new system, begins at the freezing point at 0, and places the boiling point at 100: its graduation, therefore, is decimal, and its degrees are to those of Reaumur as five to four, and to those of Fahrenheit as five to nine.

The application of the new metrology to the moneys and coins of France, has been made with considerable success; not, however, with so much of the principle of uniformity as might have been expected, had it originally formed a part of the same project. But the reformation of the coins was separately pursued, as it has been with us:

and, as the subject is of great complication, it naturally followed that, from the separate construction of two intricate systems, the adaptation of each to the other was less correct than it would have been, had all the combinations of both been included in the formation of one great master-piece of machinery. It is to be regretted that, in the formation of a system of weights and measures, while such extreme importance was attached to the discovery and assumption of a national standard of long measure as the link of connection between them all, so little consideration was given to that primitive link of connection between them, which had existed in the identity of weights and of silver coins, and of which France, as well as every other nation in Europe, could still perceive the ruins in her monetary system then existing. Her livre tournois, like the pound sterling, was a degeneracy, and a much greater one, from a pound weight of silver, but it had scarcely a seventieth part of its original value. It was divided into twenty sols or shillings; and the sol was of twelve deniers or pence. It had become a mere money of account: but the ecu, or crown, was a silver coin of six livres, nearly equivalent to an ounce in weight, and there were half crowns, and other subdivisions of it, being coins of one-fourth, one-fifth, one-eighth, and one-tenth, of the crown. There were also coins of gold, of copper, and of mixed metal called billon, in the ordinary circulations of exchange. Shortly after the adoption of the provisional or temporary metre and kilogramme, a law of 16 Vendemaire 2, (7th October, 1793,) prescribed that the principal unit, both of gold and of silver coins, should be of the weight of ten *grammes*. The proportional value of gold to silver was retained as it had long before been established in France, at $15\frac{1}{2}$ for one: the alloy of both coins was fixed at one-tenth; and the silver franc of that coinage would have been worth about thirty-eight cents, and the gold franc a little short of six dollars. This law was never carried into execution. It was superseded by one of 15th August, 1794, (28 Thermidor 3,) which reduced the silver franc to five grammes: and it was not until after a law of 7 Germinal 11, (28th March, 1803,) that gold pieces of twenty and forty francs were coined at 155 of the former to the kilogramme.

In the new system, the name of *livre*, or pound, as applied to money or coins, was discarded: but the *franc* was made the unit both of coins and moneys of account. The franc was a name which had before been in common use as a synonymous denomination of the *livre*. The new franc was of intrinsic value $\frac{1}{80}$ more than the livre. The franc is decimally divided into *decimes* of $\frac{1}{10}$, centimes of $\frac{1}{100}$, and millimes of $\frac{1}{1000}$, of the unit; but the smallest copper coin in common use is of five centimes, equivalent to about one of our cents. The silver coins are of one-fourth, one-half, one and two francs, and of five francs; the gold pieces, of twenty and forty francs. The proportional value of copper to silver is of one to forty, and that of billon to silver of one to four: so that the kilogramme should weigh 5 francs of copper coin, 50 of the billon, 200 of the silver, and 3100 of the gold coins: and the decime of billon should weigh precisely two grammes. The allowances, known by the name of remedy for errors in the weight and purity of the coins, are of $\frac{2}{100}$ upon copper, which is only for excess: those upon the weight of billon are of $\frac{14}{1000}$; upon silver $\frac{20}{1000}$ for one quarter francs, $\frac{14}{1000}$ for one-half francs, and of $\frac{10}{1000}$ or one per cent. on one and two franc pieces, and of $\frac{6}{1000}$ for five franc pieces. That of the gold coins is of $\frac{4}{1000}$; all, excepting the copper, allowances either for excess or deficiency. But the practice of the mint never transgresses in excess; and the deficiency is always nearly the whole allowed by law. The remedy of alloy is of $\frac{7}{1000}$, either of excess or defect, for billon; of $\frac{3}{1000}$ for silver; and of $\frac{2}{1000}$ for gold. It is said that the actual purity of the coins, both of gold and of silver. is within $\frac{1}{1000}$ less than the standard.

The conveniences of this system are,

First, The establishment of the same proportion of alloy to both gold and silver coins, and that proportion decimal.

Secondly, The established proportions of value between gold, silver, mixed metal, and copper coins.

Thirdly, The adaptation of all the coins to the weights in such manner as to be checks upon and tests of each other. Thus the decime of billon should weigh two grammes, the franc of silver five, the two franc piece of silver and the five centime piece of copper each ten, and the five franc piece fifty. The allowances of remedy disturb partially these proportions. These are practices continued in all the European mints, after the reasons upon which they were originally founded have in a great measure ceased. In the imperfection of the art, the mixture of the metals used in coining, and the striking of the coins, could not be effected with entire accuracy. There would be some variety in the mixture of metals made at different times, though in the same intended proportions, and in different pieces of coin, though struck by the same process and from the same die. But the art of coining metals has now attained a perfection, that such allowances have become, if not altogether, in a great measure unnecessary. Our laws make none for the deficiencies of weight: and they consider every deficiency of purity as an error, for which the officers of the mint shall be *excused* only in case of its being within $\frac{1}{143}$ part, or about $\frac{7}{1000}$; for if it should exceed that, they are disqualified from holding their offices. Where the penalty is so severe, it is proper that the allowance should be large; but, as obligatory duty upon the officers of the mint, an allowance of $\frac{1}{1000}$ would be amply sufficient for each single piece, and no allowance should be made upon the average.

Among the difficulties attending all innovations upon established usages relative to weights and measures, are their application to the tonnage of ships and boats, and to the form and size of casks. We have seen, in the review of the history of English weights and measures, how Henry the Seventh's change of the Rochelle for the troy pound affected the barrels of herring fishers, the hogshead of claret, and the butts of Alicant wine. The tonnage of ships, on the old established metrologies, was founded, like their weights and measures of capacity, upon a principle of combining specific gravity and occupied space. The ton of shipping was adapted both for a weight and a measure. The capacity of a ship as a measure is ascertained by its internal cubical dimensions, which, before the change of system in France, gave 42 royal cubic feet to a ton. The mode of admeasurement was, like ours, a complicated multiplication and division of length, breadth, and thickness, with given deductions and estimates, all finally divided by the standing number, 94, as ours is by 95, and the quotient of which gives the number of what may be called custom house *tons*. But the French ordinances, like our law, did not indicate by what specific measure this length, breadth, and thickness, were to be taken. It was always perfectly understood here, that it is in feet, and tenths of feet; and in France, that it was in royal feet and their tenths. Nothing can afford a more striking illustration of the construction which long established usage can give to law, than this admeasurement in feet and tenth parts of a foot; differing from that used in all other cases of feet and inches or twelfth parts; not expressly directed by law, and yet practised for these thirty years, probably without a question upon the meaning of the law. The attempt in France to apply it to the admeasurement by the metre, without changing the final common divisor 95, signally shows how cautiously complications of weights, measures, numbers, and coins, must be dealt with. The law of the 12th Nivose 2, (1st January, 1794,) directed those measures to be taken in the new metre and divisions, without changing the final divisor, 95, to produce a number of *tons*. The consequence would have been, that the cubic numbers divided by 95 would have been metres and their decimal parts, instead of feet and their decimal parts; and the quotient would have reduced the tonnage to about one third of its proper dimensions. To have produced a quotient of a number of tons, their final divisor should have been 30 instead of 95. This mistake was precisely the same as that of the British parliament of 1496, when, thinking to re-enact the law of 1266, they prescribed a bushel to be made from sixty-four gallons *troy* weight of wheat of thirty-two kernels to the troy pennyweight, instead of a bushel of sixty-four pounds sterling at fifteen ounces to the pound, of wheat, thirty-two kernels of which weighed the penny sterling of Henry the Third. It was the same mistake which the Greek Church yearly repeats in celebrating Easter, by the Julian calendar of 365 days 6 hours to the year, and the lunar cycle of nineteen years. And, to come nearer home to ourselves, it was the same mistake which our own statute book discloses, in estimating the British pound sterling four dollars forty-four cents, because one hundred and ten years ago Sir Isaac Newton found the Spanish Mexican piece of eight to be of the intrinsic value of four shillings and six pence sterling.

The burden of a ship, as a weight, is ascertained by the depth of the water that she draws. On the principles of hydrostatics, the weight of any floating object is equal to that of the mass of water displaced by it: and the weight of a ship's burden is the difference between the column of water drawn by her when in ballast, and when laden. The draft of water, therefore, measured by the metre and its divisions, gives of itself the result, in tons of 1000 kilogrammes, by the mere multiplication of the dimensions of the vessels; the result giving cubic metres of water, each of which, saving the difference between the specific gravity of river or sea, and distilled water, will of course be of 1000 kilogrammes.

The size of casks was among the objects intended to be included in the reformed system: and regulations were adopted prescribing, first, that their dimensions should be of uniform proportions, the diameter of the two ends, that of the centre, and the length of the barrels, being as 8, 9, and $10\frac{1}{2}$ to each other; and, secondly, that their contents should be in decimal or subdecimal divisions of *litres*. Tables were published prescribing the dimension in millimetres of the length and diametres of each cask, from the contents of 50 to those of 1000 litres. But the forms and proportions of casks are different in different countries, and in different places of the same country. These differences may arise from the nature of the substance, liquid or dry, which they are to contain; from the materials of which they are made or with which they are bound; from laws or usages long established, to which the cooper, the vintner, or brewer, the merchant, the miller, and other numerous professions dealing in articles which are packed in barrels, have accommodated themselves from time immemorial. With regard to articles of exportation, the laws of other countries also interpose, by prohibiting their admission in casks of other dimensions than those which have been used: and the instruction of 2 Frimaire 11, (23d November, 1802) revoked the regulation of Pluviose 7, (January, 1799) requiring only thenceforth, according to the proclamation of 11 Thermidor 7, (29th July, 1799,) that no wines or other liquors should be exposed to sale, unless branded with the mark of their contents in litres; with a recommendation, however, that casks should be made as much as possible in the dimensions and proportions which had been ordained in January, 1799.

The intentions of reformation upon the principles of uniformity and of decimal divisions were, in the novelty of the system, extended to the mariner's compass, which it was proposed to divide into forty rhumbs of wind, instead of thirty-two; to the log-line, the usual divisions of which are proportioned to the marine mile of sixty to a degree; to the sounding line, which had usually been divided by French mariners, not into their fathoms of six, but into *brasses* of five royal feet; and to the cable's length, which was of 100 toises. Some of these were consequences of the project for dividing decimally time, and the quadrant of the circle: and the others followed from the substitution of the metre for the foot and toise.

The lapidaries and dealers in precious stones, throughout Europe, have a weight peculiar to themselves, under the denomination of the carat, which is nearly of the weight of three grains troy, and which they divide into halves, quarters, eighths, and sixteenths. As this trade is of extremely limited extent, even in Europe, it was to be considered only in the organization of a system for universal application.

It has been observed, that, among the difficulties hitherto insuperable, which have opposed the establishment in fact of this system, thus apparently established by law, the most unmanageable of all has been found to be the adoption of the nomenclature. It is curious to observe the various expedients of legislation, to accommodate itself to the popular humors in this respect.

The law of the 1st of August, 1793, established all the principles of the new system, but under denominations different from those which had ever been used before, and not less different from those which have been adopted since. It directed the Academy of Sciences to compose an elementary book, containing a clear explanation of the new weights and measures, with tables of equalization, and instructions for adapting them to those which had been in use until then. A few days afterwards the academy was itself abolished: but the duty of composing the book was assigned to a temporary commission, or board of weights and measures, consisting of the same persons who had been employed as members of the academy on the work. The book was composed and published in the year 1794. But, on the 19th of January of that year, (30 Nivose 2,) the nomenclature had already been changed; and, on the 7th of April, 1795, (18 Germinal 3,) a nomenclature entirely new, with the exception of three or four words, was enacted. The names ordained by this law of 7th April, 1795, are still the proper technical appellations, and have already been mentioned, with their Greek and Latin prefixes of decimal multiples and subdivisions. The same law directed that weights or measures might be made of double, or of half the units and their tenth part, or tenth fold amounts; but that no other subdivision, or multiple, such as thirds, or quarters, or sixth, or eighth parts, should be allowed. The law of 19 Frimaire 8, (10th December, 1799,) declared the platina metre of 443,296 lines, and the kilogramme of 18,827.15 grains mark weight, to be the definitive standard weight and measure; on the 13th Brumaire 9, (4th November, 1800,) the executive directory issued an arrêté, or order, authorizing, either in public writings or in habitual usage, what they called a translation into French words of the authentic nomenclature; so that the myriametre might be called a league, the kilometre a mile, the litre a pint, the kilogramme a pound, the hectogramme an ounce, the gramme a denier, and so of all the rest, excepting the *metre*, which was to have no synonymous or translated name, and the *stere*, for firewood and measures of solidity. This ordinance was never executed: and the minister of the interior, by an order of 30 Frimaire 14, (21st December, 1805,) directed all the subordinate administrations to use exclusively the denominations prescribed by the law of 7th April, 1795.

An imperial decree of 12th February, 1812, presents the subject under a new aspect, by ordaining,

1. That the *units* of weights and measures should remain unchanged, as established by the law of 10th December, 1799.

2. That the minister of the interior should cause to be made *instruments* for weight and mensuration, presenting the fractions or multiples of the said units the most commonly used in commerce, and accommodated to the wants of the people.
3. That these instruments should bear on their respective faces the comparison of the divisions and denominations established by law, with those which had been formerly used.
4. That after a term of ten years a report should be made to the emperor of the result of experience upon the improvements of which the system of weights and measures might be susceptible.
5. That in the mean time the legal system should continue to be taught in all the schools, and be exclusively used in all the public offices, and in all markets, halls, and commercial transactions.

For the execution and explanation of this decree, an ordinance was, on the 28th of March, 1812, issued by the minister of the interior, of the following purport:

Art. 1. Permission was granted to employ for the purpose of commerce,

1. A long measure equal to two metres, to be called a toise, and to be divided into six feet.
2. A measure equal to one third of the metre, to be called a foot, to be divided into twelve thumbs, and the thumb into twelve lines.

Each of these measures shall bear on one side the corresponding divisions of the metre, that is to say: the toise, two metres, divided into decimetres, and the first decimetre into millimetres; and the foot, three decimetres and one third, divided into centimetres and millimetres, in all $333\frac{1}{3}$ millimetres.

Art. 2. All cloths may be measured by a stick equal in length to twelve decimetres, to be called an *ell*, (aune,) which shall be divided into halves, quarters, eighths, and sixteenths, as well as into thirds, sixths, and twelfths. It shall bear on one of its sides the corresponding divisions of the metre, in centimetres only; that is to say, one hundred and twenty centimetres, numbered from ten to ten.

Art. 4. Corn and other dry measure articles may be measured, *in sales at retail*, by a vessel equal to one-eighth of the hectolitre, which shall be called a boisseau, and shall have its double, its half, and its quarter.

Art. 5. For *retail* sales of corn, seeds, meal, and roots, green or dry, the litre may be divided into halves, quarters, and eighths.

Art. 7. For *retail* sales of wine, brandy, and other liquors, measures of one-quarter, one-eighth, and one-sixteenth of the litre may be used; each of which measures shall be called by a name signifying its proportion to the litre.

Art. 8. For retail sales of all articles which are sold by weight, the shopmen may employ the following *usual* weights:

ON WEIGHTS AND MEASURES.

The pound, (livre,) equal to half a kilogramme, or 500 grammes, which shall be divided into sixteen ounces.

The ounce, (once,) or sixteenth part of the pound, which shall be divided into eight gros.

The gros, or eighth part of the ounce, which shall be divided into halves, quarters, and eighths.

They shall bear, with their appropriate names, the indication of their weight in grammes, namely:

The pound	500 grammes
Half pound	250
Quarteron	125
Eighth, or ½ quarter	62.5
Ounce	31.3
Half ounce	15.6
Quarter ounce, 2 gros	7.8
Gros	3.9

And such is at this day the system of weights and measures, or, rather, such are the systems existing in France in their present condition; for, it cannot escape observation, that this decree and explanatory ordinance engraft upon the legal system an entirely new system, founded upon different, and, in many important respects, opposite principles. So that the result hitherto of the most stupendous and systematic effort ever made by a nation to introduce uniformity in their weights and measures, has been a conflict between four distinct systems:

1. That which existed before the Revolution.
2. The temporary system established by the law of 1st August, 1793.
3. The definitive system established by the law of 10th December, 1799. And,
4. The *usual* system, *permitted* by the decree of 12th February, 1812.

This last decree is a compromise between philosophical theory and inveterate popular habits. Retaining the principle of decimal multiplication and division for the legal system, it abandons them entirely in the weights and measures which it allows the people to use. Instead of the metre and its decimals, it gives the people a toise of six feet, an aune of three feet and one-fifth, a foot of twelve thumbs, and a thumb of twelve lines. And these measures, instead of divisions exclusively decimal, are divisible in halves, thirds, quarters, sixths, eighths, twelfths, and sixteenths. Instead of a decimated kilogramme, it gives them a pound of sixteen ounces, an ounce of eight gros, and a gros of seventy-two grains. The measures of capacity, wet and dry, have the same indulgence: and while the standard weight and measure are deposited in the national archives, the people have restored to them for use all the names and divisions of their ancient weights and measures, though not the same things. For the toise, which is twice the length of the metre, is not the old toise; the foot, which is the third part of the metre, is not the *pied de roi:* but both are longer measures. The half kilogramme, which is a pound, is not the ancient mark weight pound; nor are the boisseau or litre those of ancient times : they are all respectively near approximations to them.

If the existing system and practice terminated here, it would be far from having attained the ideal perfection of uniformity; but it is believed that, for a multitude of purposes, with this double and complicated system, there is yet a very extensive remnant in use of that which prevailed before the revolution. It appears, from questions at this time in discussion between the governments of the United States and of France, that the tonnage of the French shipping is calculated by admeasurements in cubic royal feet: and it appears hence probable that, in all the business of ship building, and in practical navigation, those measures are still used. Without positive knowledge of the fact, the analogy of all experience warrants the conjecture, that in every part of France, remote from the capital, not only the use of the old legal system, but of the local weights and measures which prevailed in the various cities and districts of the country, is far from being eradicated.

The changes which have forced themselves upon the new system, under the attempt to reduce it to practice, should serve as admonitions to correct the errors of theory; but not operate as discouragement to the pursuit of the principal object, *uniformity*. The French metrology, in the ardent and exclusive search for an universal standard from nature, seems to have viewed the subject too much with reference to the nature of things, and not enough to the nature of man. Its authors do not appear to have considered, in all the bearings of the system, the proportions dictated by nature between the physical organization of man, and the *unit* of his weights and measures. The standard taken from the admeasurement of the earth had no reference to the admeasurement and powers of the human body. The metre is a rod of forty inches: and by applying to it exclusively the principle of decimal divisions, no measure corresponding to the ancient *foot* was provided. An unit of that denomination, though of slightly varied differences of length, was in universal use among all civilized nations: and the want of it is founded in the dimensions of the human body. Perhaps for half the occasions which arise in the life of every individual for the use of a linear measure, the instrument, to suit his purposes, must be portable, and fit to be carried in his pocket. Neither the metre, the half metre, nor the decimetre, are suited to that purpose. The half metre corresponds indeed with the ancient cubit: but perhaps one of the causes which have every where, since the time of the Greeks, substituted the foot in the place of the cubit, has been the superior convenience of the shorter measure. Besides which, the cubit being the unit, the half cubit might serve the purposes of the foot; but the metre, divisible only by two and by ten, gave no measure practically corresponding with the foot whatever. It appears also not to have been considered, that decimal arithmetic, although affording great facilities for the computation of numbers, is not equally well suited for the divisions of material substances. A glance of the eye is sufficient to divide material substances into successive halves, fourths, eighths, and sixteenths. A slight attention will give thirds, sixths, and twelfths. But divisions of fifth and tenth parts are among the most difficult that can be performed without the aid of calculation. Among all its conveniences, the decimal division has the great disadvantage of being itself divisible only by the numbers two and five. The duodecimal division, divisible by two, three, four, and six, would offer so many advantages over it, that while the French theory was in contemplation, the question was discussed, whether the reformation of weights and measures should not be extended to the system of arithmetic itself, and whether the number twelve should not be substituted for ten, as the term of the periodical return to the unit. Since the establishment of the French system, this idea has been reproduced by philosophical critics, as an objection against it: and Delambre, in the third volume of the Base du Systeme Metrique, p. 302, has considered it, and assigned the reasons for which it had been rejected. He admits, to the full extent, the advantages of a duodecimal over a decimal arithmetic; but alleges the difficulty of effecting the reformation, as the decisive reason against attempting it.

The review of the proceedings in Great Britain and France, relating to the uniformity of weights and measures, presents the general subject under two very different aspects, from the combination of which, it is believed, useful practical results may be derived. Considered as a whole, the established weights and measures of England are but the ruins of a system, the decays of which have been often repaired with materials adapted neither to the proportions, nor to the principles of the original construction. The metrology of France is a new and complicated machine, formed upon principles of mathematical precision, the adaptation of which to the uses for which it was devised is yet problematical, and abiding with questionable success the test of experiment.

The standard of nature of the English system is the length of the human foot, divided by the barley corn. That of the French system is an aliquot part of the circumference of the earth decimally divided.

The material positive standard of the English system is an iron three foot rod in the British exchequer. That of France is a platina metre in the national archives.

To the English system belong two different units of weight, and two corresponding measures of capacity, the natural standard of which is the difference between the specific gravities of wheat and wine. To the French system there is only one unit of weight and one measure of capacity, the natural standard of which is the specific gravity of water.

The French system has the advantage of unity in the weight and the measure, but has no common test of both. Its measure gives the weight only of water. The English system has the inconvenience of two weights and two measures; but each measure is at the same time a weight. Thus the gallon of wheat and the gallon of wine, though of different dimensions, balance each other as weights. A gallon of wheat and a gallon of wine, each, weigh eight pounds avoirdupois. This observation applies, however, only to the original principle of the English system, and not altogether to its present condition. The difference between the specific gravity of wheat and wine, is still the difference between the troy and avoirdupois weights, but not between the wine and corn gallons. A third vessel of capacity, for which neither the necessity nor the use is perceived, has usurped the place of the corn gallon; and it has been shown how it was introduced. The acts of parliament prescribing the dimensions of the bushel and of the wine gallon in cubic inches, have assumed them from existing standards, or erroneous calculations: and the proportions between the measures of corn and of wine, which belonged originally to the system, are now transferred to those of wine and beer, for which, if the reason was that beer being a home made liquor and wine a foreign production, beer a comfort of the poor, and wine a luxury of the rich, the former ought to be dealt out in larger portions, and more lightly touched with taxation, it proceeded from the best motives of political morality; but which might have been as well accomplished by reducing the tax as by enlarging the measure. As vessels of capacity for fluids, there can be no useful reason for different measure, except the proportion of specific gravities.

In the English system, the smaller of the two weights was originally also identical with the coin: a pound of the weight was a pound sterling in silver money. But this property it has irrecoverably lost.

In the French system, the weight is not a coin; but the metallic coins are weights. Gold, silver, mixed metal, and copper, are all coined in proportions of weight and relative value prescribed by law.

In our monetary system we have discarded the last trace of identity between weights and coins, by ceasing to apply to money the name of pound or penny. Our coins are of prescribed weight and purity, but in no convenient or uniform proportions to each other.

In the English system the two weights are standards of verification to each other; the two pounds being in the proportion to each other of 144 to 175, and the pound avoirdupois being of 7,000 grains troy. For quantities amounting to one fourth of a hundred pounds or more, the English avoirdupois weight requires an accession of 12 per cent.; 28 pounds pass for 25, 56 for 50, and 112 for 100. The original motive for this must have been the convenience of dividing the hundred into halves, quarters, eighths, and sixteenths, without making fractions of a pound. The true hundred can thus be divided into no whole number less than a quarter, or 25.

In the English system, the standard linear measure is connected with the weights by the specific gravity of spring water, of which a measure of one cubic foot contains one thousand ounces avoirdupois.

In the French system, the standard linear measure is connected with the weight and the measure of capacity, by the specific gravity of distilled water, at its greatest density, one cubic decimetre of such water being the weight of the kilogramme, and filling the measure of the litre.

In the English system, every weight and every measure is divided by different and, seemingly, arbitrary numbers; the foot into twelve inches; the inch, by law, into three barley corns, in practice sometimes into halves, quarters, and eighths, sometimes into decimal parts, and sometimes into twelve lines; the pound avoirdupois into sixteen ounces, and the pound troy into twelve, so that while the pound avoirdupois is heavier, its ounce is lighter than those of the troy weight. The ton, in the English system, is both a weight and a measure. As a measure, it is divided into four quarters, the quarter into eight bushels, the bushel into four pecks, &c. As a weight, it is divided into twenty hundreds, of 112 pounds, or 2240 pounds avoirdupois. The gallon is divided into four quarts, the quart into two pints, and the pint into four gills.

In the French system, decimal divisions were prescribed by law exclusively. The binary division was allowed, as being compatible with it: but all others were rigorously excluded; no thirds, no fourths, no sixths, no eighths, or twelfths. But this part of the system has been abandoned: and the people are now allowed all the ancient varieties of multiplication and division, which are still further complicated by the decimal proportions of the law.

The nomenclature of the English system is full of confusion and absurdity, chiefly arising from the use of the same names to signify different things; the term pound to signify two different weights, a money of account, and a coin; the gallon and quart to signify three different measures; and other improper denominations constantly opening avenues to fraud.

The French nomenclature possesses uniformity in perfection, every word expressing the unit weight or measure which it represents, or the particular multiple or division of it. No two words express the same thing: no two things are signified by the same word.

If, with a view to fixing the standard of weights and measures for the United States, upon the principles of the most extensive uniformity, the question before Congress should be upon the alternative, either to adhere to the system which we possess, or to adopt that of France in its stead, the first position which occurs as unquestionable is, that change, being itself diversity, and therefore the opposite of uniformity, cannot be a means of obtaining it, unless some great and transcendent superiority should demonstrably belong to the new system to be adopted, over the old one to be relinquished.

In what then does the superiority of the French system, in all its novelty and freshness, over that of England, in all its decays, theoretically consist?

1. In an invariable standard of linear measure, taken from nature, and being an aliquot decimal portion of the quarter of the meridian.

2. In having a single unit of all weights, and a single unit of measures of capacity for all substances, liquid or dry.

3. In the universal application of the decimal arithmetic, to the multiples and divisions of all weights and measures.

4. In the convenient proportions by which the coins and money of account are adjusted to each other and to the weights.

5. In the uniformity, precision, and significancy, of the nomenclature.

1. If the project of reforming weights and measures had extended, as was proposed by the French system, to the operations of astronomy, geography, and navigation; if the quadrant of the circle and of the sphere had been divided into one hundred degrees, each of one hundred thousand metres; the assumption of that measure would have been an advantage much more important than it is, or can be, in the present condition of the system. Whether it would have compensated for disturbing that uniformity which exists, and which has invariably existed, of the division into ninety degrees, with sexagesimal subdivisions of minutes and seconds, is merely matter of speculation. At least, it has been found impracticable, even in France, to carry it into effect: and, without it, the metre, as the natural standard of the system, has no sensible advantage over the foot. To a perfect system of uniformity for all weights and measures, an aliquot part of the circumference of the earth is not only a better natural standard unit than the pendulum, or the foot, but it is the only one that could be assumed. Every voyage round the earth is an actual mensuration of its circumference. All navigation is admeasurement: and no perfect theory of weights and measures could be devised, combining in it the principle of decimal computation, of which any other natural standard whatever could accomplish the purpose. Its advantages over the pendulum are palpable. The pendulum bears no proportion to the circumference of the earth, and cannot serve as a standard unit for measuring it. Yet a system of weights and measures, which excludes all geography, astronomy, and navigation, from its consideration, must be essentially defective in the principle of uniformity.

But, if the metre and its decimal divisions are not to be applied to those operations of man, for which it is most especially adapted; if those who circumnavigate the globe in fact are to make no use of it, and to have no concern in its proportions; if their measures are still to be the nonagesimal degree, the marine league, the toise, and the foot; it is surely of little consequence to the farmer who needs a measure for his corn, to the mechanic who builds a house, or to the townsman who buys a pound of meat, or a bottle of wine, to know that the weight, or the measure which he employs, was standarded by the circumference of the globe. For all the uses of weights and measures, in their ordinary application to agriculture, traffic, and the mechanic arts, it is perfectly immaterial what the natural standard, to which they are referable, was. The foot of Hercules, the arm of Henry the First, or the barley-corn, are as sufficient for the purpose as the pendulum, or the quadrant of the meridian. The important question to them is, the correspondence of their weight or measure with the positive standard. With the standard of nature, from which it is taken, they have no concern, unless they can recur to it as a test of verification. However imperfect for this end the human foot, or the kernel of wheat or barley, may be, they are at least easily accessible. It is a great and important defect of the systems which assume the meridian or the pendulum for their natural standard, that they never can be recurred to without scientific operations.

This is one great advantage which a natural standard, taken from the dimensions and proportions of the human body, has over all others. We are perhaps not aware how often every individual, whose concerns in life require the constant use of long measures, makes his own person his natural standard, nor how habitually he recurs to it. But the habits of every individual inure him to the comparison of the definite portion of his person, with the existing standard measures to which he is accustomed. There are few English men or women but could give a yard, foot, or inch measure, from their own arms, hands, or fingers, with great accuracy. But they could not give the metre or decimetre, although they should know their dimensions as well as those of the yard and foot. When the Russian General Suwarrow, in his Discourses under the Trigger, said to his troops, "a soldier's step is an arsheen;" he gave every man in the Russian army the natural standard of the long measure of his country. No Russian soldier could ever afterwards be at a loss for an arsheen. But, although it is precisely twenty-eight English inches, being otherwise divided, a Russian soldier would not, without calculation, be able to tell the length of an English yard or inch.

Should the metre be substituted as the standard of our weights and measures, instead of the foot and inch, the natural standard which every man carries with him in his own person would be taken away; and the inconvenience of the want of it would be so sensibly felt, that it would be as soon as possible adapted to the new measures: every man would find the proportions in his own body corresponding to the metre, decimetre, and centimetre, and habituate himself to them as well as he could. If this conjecture be correct, is it not a reason for adhering to that system which was founded upon those proportions, rather than resort to another, which, after all, will bring us back to the standard of nature in ourselves.

2. The advantage of having a single unit of all weights and a single unit of measures of capacity, is so fascinating to a superficial view, that it would almost seem presumption to raise a question, whether it be so great as at first sight it appears. The relative value of all the articles which are bought and sold by measures of capacity, is a complicated estimate of their specific gravity and of the space which they occupy. If both these properties are ascertained by one instrument for any one article, it cannot be applied with the same effect to another. Thus the litre, in the French system, is a measure for all grains and all liquids: but its capacity gives a weight only for distilled water. As a measure of corn, of wine, or of oil, it gives the space which they occupy, but not their weight. Now, as the weight

of those articles is quite as important in the estimate of their quantities as the space which they fill, a system which has two standard units for measures of capacity, but of which each measure gives the same weight of the respective articles, is quite as uniform as that which, of any given article, requires two instruments to show its quantity; one to measure the space it fills, and another for its weight. It has been observed, that nature, in the relations which she has established between man and the earth upon which he dwells, and in providing for the wants resulting to him from these relations, offers him in his own person two natural standards even of linear measure; one for the range of his own movements upon the earth, and the other for articles loosened from the earth, and which are adapted to the immediate wants of his person. He finds by experience that these may with increased convenience be reduced to one. It is not exactly so with weights or measures of capacity. From the moment when man becomes a tiller of the ground, and civil society is organized; from the moment when the mutual exchange between the wants of one and the superfluities of another commences; measures of capacity and weights are necessary to the operation. The use of metals, as common standards of value, is of later origin, and, when first applied to that purpose, they are always delivered by weight. The first and most important article of traffic is corn, the first necessary of life: wine and oil successively come next: milk and honey follow. For all these, weights and measures of capacity are indispensable. When the metals are first used as common instruments of exchange, the proportions of their qualities are estimated by their weight. But that weight could not be ascertained by itself. The metal being in one scale, there must be something else to balance it in the other: and that other substance, first of all, would, whenever it should have come into use for food, be corn. It might next be wine. But thus compared, it would immediately be seen that the vessel containing of wine a counterpoise to the given metallic weight, would not contain a counterpoise of wheat to the same weight: and what could more naturally suggest itself than the device, to bring to the scales the wheat in a measure to balance the weight, and the wine in a measure to produce the same effect? The metallic weight would then become the common standard for both, but would neither be the same weight by which its own gravity had been ascertained, nor a substitute for it. Thus, the operation of weighing implies in its nature the use of two articles, each of which is the standard testing the gravity of the other. And in the difference between the specific gravities of corn and wine, nature has also dictated two standard measures of capacity, each of them equiponderant to the same weight.

This diversity existing in nature, the troy and avoirdupois weights, and the corn and wine measures of the English system are founded upon it. In England it has existed as long as any recorded existence of man upon the island. But the system did not originate there, neither was Charlemagne the author of it. The weights and measures of Rome and of Greece were founded upon it. The Romans had the *mina* and the *libra*, the nummulary pound of twelve ounces, and the commercial pound of sixteen. And the Greeks, as well as the Romans, had a weight for small and precious, and a weight for bulky and cheap commodities. The Greeks denominated them by significant terms, *the weight for measure,* and *the weight for money.* Whether the ounce, of which these pounds were composed, was the same, is a subject of much controversy, but of little importance to decide. At the period of the lower empire, these two weights were known by the name of the *eastern* and *western* pound. And the denomination of the former was the same in England: it was the *easterling* pound, and the origin of the term sterling in the English language: it was the pound of the eastern nations, by which Europe was overrun in the decline of the Roman Empire. The avoirdupois pound had the same origin: for it came through the Romans from the Greeks, and through them, in all probability, from Egypt. Of this there is internal evidence in the weights themselves, and in the remarkable coincidence between the cubic foot and the thousand ounces avoirdupois, and between the ounce avoirdupois and the Jewish silver shekel. The Greek foot was, within a fraction of less than the hundredth part of an inch, the same with that of England. The ounce avoirdupois is the same with the Roman and Attic ounce, and the exact double of the Jewish shekel. The Silian plebiscitum, or ordinance of the Roman people, of the year 509, two hundred and fifty years before the Christian era, declares, that a quadrantal of wine shall be eighty pounds, a congius of wine ten pounds; that six sextarii make a congius of wine, forty-eight sextarii a quadrantal of wine; that the sextarius of liquid and dry measures should be the same; and that sixteen pounds make the modius. The congius was the Roman gallon, and the modius the Roman peck. The quadrantal was the same as the amphora, and was formed from the cubic foot of water, so that eighty pounds of wine were equal to a cubic foot of water.

The same combinations are traced with equal certainty to the Greeks and Egyptians: and, if the shekel of Abraham was the same as that of his descendants, the avoirdupois ounce may, like the cubit, have originated before the flood.

This diversity is, therefore, founded in the nature of things; and may be stated by the following rule: that whatever is sold *by* weight, in *measure* must have a measure for itself, which will serve for no other article, of different specific gravity; and as wheat and wine are both articles of that description, as their specific gravities are very materially different, although they are very suitable to be weighed by the same weight, they yet require different measures, to place them in equipoise with that weight. The difference of specific gravity between the vinous and watery fluids is so slight, that neither in the Greek, the Roman, nor the English system, was there any account taken of it. But with regard to oil, it appears that the Greeks had a separate measure adapted to its specific gravity, which they considered as being in proportion to that of wine or water as nine to ten.

Notwithstanding, therefore, the first appearance of superior uniformity and simplicity, presented by the single unit of weights, and single measure of capacity in the new system of France, it appears to be more conformable to the order of nature, and more subservient to the purposes of man, that there should be two scales of weight and two measures of capacity, graduated upon the respective specific gravities of wheat and wine, than with a single weight and a single measure, to be destitute of any indication of weight in the measure.

This conclusion has been confirmed by a very striking fact, which has occurred in France under the new system. By an ordinance of police, approved on the 6th of December, 1808, by the Minister of the Interior, it is prescribed that the sale of oil in Paris by retail shall be *by* weight, in measures, containing five hectogrammes, one double hectogramme, one hectogramme, &c. And these measures, being cylinders of tin, are stamped with initial letters, indicating that one is for sweet oil, and the other for lamp oil. So that here are two new measures of capacity altogether incongruous to the new system, each differing in cubic dimensions from the other, though to measure the single article of oil, and both differing from the litre. They attach themselves indeed to the new system by *weight*, but abandon entirely its pretensions to unity of measure; and fall at once into the principle of the old system, of adapting the measure to the weight.

By the usages of modern times, the weight of wine is of little or no consideration. Its first admeasurement is in casks, of different dimensions in different places, and which cannot be made uniform, unless by a system of metrology common to many nations. It is sold wholesale by the cask or hogshead, the contents of which are ascertained by mechanical gauging instruments, adapted to the smaller measures of capacity of the country where it is to be consumed. These instruments give the solid contents of the vessel, and the number of the standard measures of the country which it contains. The gauging rods used in England and the United States give the contents in cubic inches and wine gallons. As a test of the quantity of wine contained in the cask, this mode of admeasurement is less certain and effectual than weight, especially if the cask is not full: but, being more convenient and easy of application, and specially adapted to the legal measure of the gallon in cubic inches, it has superseded altogether the use of weights as proofs of the quantity of wine. By retail, the article is sold either in the gallon measures fixed by law at 231 cubic inches, or in bottles of no definite measure, but containing an approximation to a quart or pint.

Our system of weights and measures, by the substitution of the wine gallon of 231 for that of 224 cubic inches, has lost the advantage which it originally possessed of testing the accuracy of a wine measure by its weight. The average specific gravity of wine is of 250 grains troy weight to a cubic inch : four inches therefore make a thousand grains, and twenty-eight inches a pint weighing one pound avoirdupois. These coincidences would be of great utility and convenience, and would be rendered still more so by another, which is, that this number of 224 inches is the exact decimal part of 2240, the number of pounds avoirdupois that go to a ton. As it now exists, therefore, the measure of the gallon of wine does not show its weight; and the unity of the measure of capacity in the French system, is an advantage not compensated by any benefit derived from the different dimensions of our corn and wine gallons.

Our country is not as yet a land of vineyards. We have no "flowery dales of Sihma, clad with vines." Wine is an article of importation; an article of luxury, in a great measure confined to the consumption of the rich. Its distribution in measure, and the exactness of the measure by which it is distributed, is not an incident which every day comes home to the interests and necessities of every individual. We have less reason for regretting, therefore, the loss of a measure which would prove its integrity by its weight; and more reason for preferring the uniformity of singleness in the French system of capacious measures, to the uniformity of proportion which belonged originally to the English. That proportion itself we have lost by the establishment of a wine gallon of 231, and a corn bushel of 2150 cubic inches: and although it exists in the troy and avoirdupois weights, and in the wine and beer gallons, it exists to none of the useful purposes for which it was originally intended, and to which in former days it was turned.

The consumption of wine in modern times is exceedingly diminished, not only by the substitution of beer, and of spirits distilled from grain in the countries where the vine is not cultivated, but by the use, now become universal, of decoctions from aromatic herbs and berries. Tea and coffee are potations unknown to the European world until within these two centuries: and they have probably diminished by one-half the consumption of wine throughout the world.

The measures, by which solid and liquid substances are sold, are not, and cannot conveniently be the same. The form and the substance of the vessels in which they are kept are altogether different. Grain is usually kept in bags, until ground into meal. Liquids, in large quantities, are kept in wooden vessels of peculiar construction, founded upon the properties of fluids and the laws of hydrostatics: in small quantities, they are kept in vessels of glass, adapted by their form to the facility of pouring them off without loss. Such vessels are utterly unsuitable for containing grain, or any other solid substance. The forms, both of casks and of bottles, are among the most difficult forms into which cubical extension can be moulded for ascertaining quantity by linear measure. They not only contain the problem, hitherto unsolvable to man, of squaring the circle; but some of the most recondite mysteries of the conic sections. They are neither cylinders, nor ellipses, nor cones, nor spheres; but a combination of all these forms. Grain may be measured by a cylindrical or a cubical vessel, at pleasure. The cylindrical form is best adapted to convenience; and by the known proportion of the diameter to the circle, its solid contents in linear measure may be ascertained with sufficient accuracy and little difficulty. Grain cannot be kept in vessels with large bodies, long necks, and narrow mouths. Liquids can be well kept for preservation in no other. Grain is a swiftly perishable substance, which must generally be consumed within a year from its growth : wines and spirituous liquors in general may be kept many years, and the vessels in which they are kept must be of forms and substances calculated to guard against loss by evaporation, fermentation, or transudation. So different indeed are all the properties of grain and of all liquids, that, instead of requiring the same measure to indicate their qualities, the call of nature is for different vessels, of different substances, and in different forms. The most certain and convenient test for the accuracy of dry measures is linear measure; that of liquids is *weight*. The sextarius of the Roman system, and the litre of the French, were measures common both to wet and dry substances. But in applying it, the Romans formed a liquid measure of ten pounds *weight*, and a dry measure of sixteen. The French litre combines both the tests of linear measure and of weight for the single article of distilled water, at a certain temperature of the atmosphere : but it is not the test of weight for any thing else. The hectolitre of wine or of corn is no indication of the weight of either. The sale of wheat, from the nature of the article, must usually be in large quantity, seldom less than a bushel. The unit of the measure declared in Magna Charta is the quarter of a ton, or eight bushels. Wine is an article the sale of which is as frequent in retail as by wholesale. The accuracy of its admeasurement in small quantities is important. In this respect it has an analogy to the precious metals. In fine, the purchase and sale of liquid and dry substances is, by the constitution of human society, not at the same times, or places, nor by the same persons. Their difference in the origin is that of the vineyard and the cornfield. They pass thence respectively to the wine press and the flour mill; thence to the vintner and the flour merchant, in vessels already adapted to their respective conditions; the corn having undergone a transformation requiring a different measure from that of wheat. Trace them through all their meanderings in the circulation of civil society, till they come to their common ultimate use for the subsistence of man; it will never be found that the same measures are necessary for, or suitable to, them. The wheat comes in the shape of meal or of bread, to be measured by weight; and the liquors in casks or bottles, and still in the form given to them by distillation. The distiller and the brewer, who manufacture the liquid from the grain, have occasion for both measures; but the articles come to them in one form, and go from them in the other; nor is there any apparent necessity that they should receive and issue them by the same measure.

There are conveniences in the intercourse of society, connected with the use of smaller and more minutely perfect weights and measures of capacity, for sales of articles by retail, than by wholesale, and for articles of great price though of small bulk. Thus, drugs, as articles of commerce, and in gross, are sold by the avoirdupois or commercial pound; used as medicines, in minute quantities, and compounded by the apothecary, they are sold by the smaller or nummulary weight. The laws of Pennsylvania authorize innkeepers to sell beer, *within the house,* by the wine measure; but, for that which they send out of the house, require them to use the beer gallon or quart. In both these cases the difference of the measure forms part of the compensation for the labor and skill of the apothecary, and part of the profits necessary to support the establishment of the publican. There is, finally, an important advantage in the establishment of two units of weights and of measures of capacity, by the possession in each of a standard for the verification of the other. It serves as a guard against the loss or destruction of the positive standard of either. The troy and avoirdupois pounds are to each other as 5,760 to 7,000. Should either of these standard pounds be lost, the other would supply the means of restoring it. The same thing might be effected by the measures of beer and of wine. The French system has designated the pendulum as such a standard for the verification of the metre. The English system gives, in each weight and measure, a standard for the other.

The result of these reflections is, that the uniformity of nature for ascertaining the quantities of all substances, both by gravity and by occupied space, is a uniformity of proportion, and not of identity; that, instead of one weight and one measure, it requires two units of each, *proportioned* to each other; and that the original English system of metrology, possessing two such weights, and two such measures, is better adapted to the only uniformity applicable to the subject, recognized by nature, than the new French system, which, possessing only one weight and one measure of capacity, identifies weight and measure only for the single article of distilled water; the English uniformity being relative to the *things* weighed and measured, and the French only to the instruments used for weight and mensuration.

3. The advantages of the English system might, however, be with ease adapted to that of France, but for the exclusive application in the latter of the decimal arithmetic to all its multiples and subdivisions. The decimal numbers, applied to the French weights and measures, form one of its highest theoretic excellencies. It has, however, been proved by the most decisive experience in France, that they are not adequate to the wants of man in society: and, for all the purposes of retail trade, they have been formally abandoned. The convenience of decimal arithmetic is in its nature merely a convenience of calculation: it belongs essentially to the keeping of accounts; but is merely an incident to the transactions of trade. It is applied, therefore, with unquestionable advantage, to moneys of account, as we have done: yet, even in our application of it to the *coins,* we have not only found it inadequate, but in some respects inconvenient. The divisions of the Spanish dollar, as a coin, are not only into tenths, but into halves, quarters, fifths, eighths, sixteenths, and twentieths. We have the halves, quarters, and twentieths, and might have the fifths; but the eighth makes a fraction of the cent, and the sixteenth

even a fraction of a mill. These eighths and sixteenths form a very considerable proportion of our metallic currency: and although the eighth dividing the cent only into halves adapts itself without inconvenience to the system, the fraction of the sixteenth is not so tractable; and in its circulation, as small change, it passes for six cents, though its value is six and a quarter, and there is a loss by its circulation of four per cent, between the buyer and the seller. For all the transactions of retail trade, the eighth and sixteenth of a dollar are among the most useful and convenient of our coins: and, although we have never coined them ourselves, we should have felt the want of them, if they had not been supplied to us from the coinage of Spain.

This illustration, from our own experience, of the modification with which decimal arithmetic is adaptable even to money, its most intimate and congenial natural relative, will disclose to our view the causes which limit the exclusive application of decimal arithmetic to *numbers*, and admit only a partial and qualified application of them to weight or measure.

It has already been remarked, that the only apparent advantage of substituting an aliquot part of the circumference of the earth, instead of a definite portion of the human body, for the natural unit of linear measure, is, that it forms a basis for a system embracing *all* the objects of human mensuration; and that its usefulness depends upon its application to geography and astronomy, and particularly to the division of the quadrant of the meridian into centesimal degrees. In the novelty of the system, this was attempted in France, as well as the decimal divisions of time, and of the rhumbs of the wind. A French navigator, suffering practically under the attempt thus to navigate, decimally, the ocean, recommended to the national assembly to decree, that the earth should perform four hundred revolutions in a year. The application of decimal divisions to time, the circle, and the sphere, are abandoned even in France. And for all the ordinary purposes of mensuration, excepting itinerary measure, the metre is too long for a standard unit of nature. It was a unit most especially inconvenient as a substitute for the foot, a measure to which, with trifling variations of length, all the European nations and their descendants were accustomed. The foot rule has a property very important to all the mechanical professions, which have constant occasion for its use: it is light, and easily portable about the person. The metre, very suitable for a staff, or for measuring any portion of the earth, has not the property of being portable about the person: and, for all the professions concerned in ship or house building, and for all who have occasion to use mathematical instruments, it is quite unsuitable. It serves perfectly well as a substitute for the yard or ell, the fathom or perch; but not for the *foot*. This inconvenience, great in itself, is made irreparable when combined with the exclusive principle of decimal divisions. The union of the metre, and of decimal arithmetic, rejected all compromise with the foot. There was no legitimate extension of matter intermediate between the ell and

ON WEIGHTS AND MEASURES. 83

the palm, between forty inches and four. This decimal despotism was found too arbitrary for endurance; not only the foot, but its duodecimal divisions, were found to be no arbitrary or capricious institutions, but founded in the nature of the relations between man and things. The duodecimal division gives equal aliquot parts of the unit, of two, three, four, and six. By giving the third and the fourth, it indirectly gives the eighth and sixteenth, and gives facility for ascertaining the ninth, or third of the third. Decimal division, in giving the half, does not even give the quarter, but by multiplication of the subdivisions. It is incommensurable with the *third*, which unfortunately happened to be the foot, the universal standard unit of the old metrology. The choice of the kilogramme, or cubical decimetre of distilled water, as the single standard unit of weights, with the application to it of the decimal divisions, was followed by similar inconveniences. The pound weight should be a specific gravity easily portable about the person, not only for the convenience of using it as an instrument, but as the measure of quantities to be carried. To the common mass of the people, the use of weights is in the market, or the shop. The article weighed is to be carried home. It is an article of food for the daily subsistence of the individual or his family. As he has not the means of purchasing it in large quantities, it must often be sold in quantities represented by the pound weight, which, like the foot rule, with various modifications, is universally used throughout the European world. Subdivisions of that weight, the half, the quarter of a pound, are often necessary to conciliate the wants and the means of the neediest portion of the people; that portion to whom the *justice* of weight and measure is a necessary of life, and to whom it is one of the most sacred duties of the legislator to secure that justice, so far as it can be secured by the operation of human institutions. The half of the kilogramme was nearly equivalent to the ancient Paris pound. But there was in the new system no half, or quarter of a pound; because there was no quarter or eighth of a kilogramme. There was no intermediate weight between the pound, or half kilogramme, and the hectogramme, which was a fifth part of a pound.

The *litre*, or unit of measures of capacity in the new system, had one great advantage over the linear and weight units, by its near equivalence to the old Paris pint, of which it was to take the place. But, on the other hand, decimal divisions are still more inapplicable to measures of capacity for liquids, than to linear measures or weights. The substance in nature best suited for a retail measure of liquids, is tin: and the best form, in which the measure can be moulded, is a slight approach from the cylinder to the cone. Our quart and gallon wine and beer measures are accordingly of that form, as are all the most ordinary vessels used for drinking. In the new French system, the form of all the measures of capacity is cylindrical; and the litre is a measure, the diameter of which is half its depth. It is, therefore, easily divisible into halves, quarters, and eighths; for it needs only thus to divide the depth, retaining the same diameter. But all conveniences of proportion are lost, by taking one-tenth of the depth and retaining the same diameter: and, if the diameter be reduced, there is no means, other than complicated calculation, squarings of the circle, and extractions of cube roots, that will give one liquid measure which shall be the tenth part of another.

In the promiscuous use of the old weights and measures and the new, which was unavoidable in the transition from the one to the other, the approximation to each other of the quarter and the fifth parts of the unit became a frequent source of the most pernicious frauds; frauds upon the scanty pittance of the poor. The small dealers in groceries and liquors, and marketmen, gave the people the fifth of a kilogramme for a half pound, and the fifth of the litre for a half setier. The most easy and natural divisions of liquids are in continual halvings: and the Paris pint was thus divided into halves, quarters, eighths, sixteenths, and thirty-second parts, by the name of chopines, half setiers, possons, half possons, and roquilles. The half setier, just equivalent to our half pint, was the measure in most common use for supplying the daily necessities of the poor; and thus the decimal divisions of the law became snares to the honesty of the seller, and cheats upon the wants of the buyer.

Thus then it has been proved, by the test of experience, that the principle of decimal divisions can be applied only with many qualifications to any general system of metrology; that its natural application is only to numbers; and that time, space, gravity, and extension, inflexibly reject its sway. The new metrology of France, after trying it in its most universal theoretical application, has been compelled to renounce it for all the measures of astronomy, geography, navigation, time, the circle, and the sphere; to modify it even for superficial and cubical linear measure, and to compound with vulgar fractions, in the most ordinary and daily uses of all its weights and all its measures. It has restored the foot, the pound, and the pint, with all their old subdivisions, though not exactly with their old dimensions. The foot, with its duodecimal divisions into thumbs and lines, returns in the form the most irreconcileable possible, with the decimals of the metre; for it comes in the proportion of three to ten, and consists of $333\frac{1}{3}$ millimetres. This indulgence to linear measure is without qualification, and may be used in all commerce, whether of wholesale or retail. The restoration of the pound, the boisseau,* and the pint, is limited to retail trade. The fractions of the pound are as averse to decimal combinations as those of the foot. The eighth of a pound, for instance, is 625 decigrammes, each of about $1\frac{1}{2}$

---

* One of the most abundant sources of error and confusion, in relation to weights and measures, arises from mistranslation of those of one country into the language of another. Thus, to call the *pinte* of Paris, a pint, is to give an incorrect idea of its contents. The Paris pinte corresponded with our wine quart, containing 46.95 French, or 58 08 English cubic inches. To call the boisseau a *bushel*, is a still greater incongruity between the word and the idea connected with it. The *boisseau* contained 655 French cubic inches, and was less than $1\frac{1}{2}$ peck English. The minot, or three boisseaus, was the measure corresponding with the English bushel.

grain troy weight. The half of this eighth is an *ounce,* to form which decimally requires a recourse to another fractional stage, and to say 31.25 milligrammes. But the milligramme, being equivalent to less than $\frac{1}{6}$ of a grain troy weight, is too minute for accurate application; so that it is called, and marked upon the weight itself as 31.3decigrammes. The half ounce, instead of 15.625 decimilligrammes, is marked for 15.6 decigrammes. The quarter of an ounce, instead of 7.8125, passes for 7.8 decigrammes, and the gros, or groat, instead of 3.90625, is abridged to 3.9. The ounce and all the smaller weights, therefore, reject the coalition of subdivision by decimal and vulgar fractions: and the weights for account are different from the weights for trade.

From the verdict of experience, therefore, it is doubtful whether the advantage to be obtained by any attempt to apply decimal arithmetic to weights and measures, would ever compensate for the increase of diversity which is the unavoidable consequence of change. Decimal arithmetic is a contrivance of man for computing numbers; and not a property of time, space, or matter. Nature has no partialities for the number ten: and the attempt to shackle her freedom with them, will for ever prove abortive.

The imperial decree of March, 1812, by the reservation of a purpose to revise the whole system of the new metrology, after a further interval of ten years of experience, seems to indicate a doubt, whether the system itself can be maintained. Ten years from 1812, was a period far beyond that which Providence had allotted to the continuance of the imperial government itself. The royal government of France, which has since succeeded, has hitherto made no change in the system. Whether, at the expiration of the ten years, limited in the decree, the proposed revisal of it will be accomplished by the present government, is not ascertained. In the mean time, the whole system must be considered as an experiment upon trial even in France: and should it ultimately prove, by its fruits, worthy of the adoption of other nations, it will at least be expedient to postpone engrafting the scion, until the character of the tree shall have been tested, in its native soil, by its fruits.

4. The fourth advantage of the French metrology over that which we possess, consists in the convenient proportions, by which the coins and moneys of account are adjusted to each other, and to the weights.

This is believed to be a great and solid advantage; not possessed exclusively by the French system, for it was, in high perfection, a part of the original English system of weights and measures, as has already been shown. It was more perfect in that system, because the silver coins and weights were not merely proportioned to each other, but the *same.* This is not the case with the French coins: and even their *proportions* to the weights are disturbed and unhinged by the mint allowance, or what they call toleration of inaccuracy, both of weight and alloy. This toleration, which is also technically called the *remedy,* ought every where to be exploded. It is in no case necessary. The toleration is injustice: the remedy is disease. If it were the duty of this report to present a system of weights, measures, and coins, all referable to a single standard, combining with it, as far as possible, the decimal arithmetic, and of which uniformity should be the pervading principle, without regard to existing usages, it would propose a silver coin of nine parts pure and one of alloy : of thickness equal to one-tenth part of its diameter; the diameter to be one-tenth part of a foot, and the foot one-fourth part of the French metre. This dollar should be the unit of weights as well as of coins and of accounts ; and all its divisions and multiples should be decimal. The unit of measures of capacity should be a vessel containing the weight of ten dollars of distilled water, at the temperature of ten degrees of the centigrade thermometer : and the cubical dimensions of this vessel should be ascertained by the weight of its contents ; the decimal arithmetic should apply to its weight, and convenient vulgar fractions to its cubical measure. This system once established, the standard weight and purity of the coin should be made an article of the constitution, and declared unalterable by the legislature. The advantage of such a system would be to embrace and establish a principle of uniformity with reference to time, which the French metrology does not possess. The weight would be a perpetual guard upon the purity and value of the coin. No second weight would be necessary or desirable. The coin and the weight would be mutual standards for each other ; accessible, at all times, to every individual. Should the effect of such a system only be, as its tendency certainly would be, to deprive the legislative authority of the power to debase the coins, it would cut up by the roots one of the most pernicious practices that ever afflicted man in civil society. By its connection of the linear standard with the French metre, it would possess all the advantages of having that for a unit of its measures of length, and a link of the most useful uniformity with the whole French metrology.

But the consideration of the coins is beyond the scope of the resolutions of the two Houses; nor is their relation to the weights and measures of the country viewed, by the constitution and laws of the United States, as that of parts of one entire system. Excepting the application of decimal divisions to our money of account, and the establishment of the dollar as the unit both of the money of account and of the silver coins, our moneys have no uniform or convenient adjustment to our weights. The proportion of alloy is not the same in our coins of silver, as in those of gold : and the only connection between our monetary system and our weights and measures is, that the gravity and proportional purity of the coins is prescribed in troy weight grains. To obtain, therefore, the advantage existing in the French metrology, of easy proportions between the weights and coins, or the still greater advantage of identity between them which belonged to the old English system, an entire change would be necessary in the fabrication of our coins, and in our moneys of account.

It is, at least, extremely doubtful whether the benefits to be derived from such a change would be equivalent to the difficulties of achieving it, and the hazard of failing in the attempt.

5. The last superior advantage of the French metrology is, the *uniformity*, precision, and significancy, of its nomenclature.

In mere speculative theory, so great and unequivocal is this advantage, that it would furnish one of the most powerful arguments for adopting the whole system to which it belongs. In every system of weights and measures, ancient or modern, with which we are acquainted, until the new system of France, the poverty and imperfection of language has entangled the subject in a snarl of inextricable confusion. The original *names* of all the units of weights and measures have been improper applications of the substances from which they were derived. Thus, the foot, the palm, the span, the digit, the thumb, and the nail, have been, as measures, improperly so called, for the several parts of the human body, with the length of which they corresponded. Instead of a specific name, the measure usurped that of the standard from which it was taken. Had the foot rule been unalterable, the inconvenience of its improper appellation might have been slight. But, in the lapse of ages, and the revolutions of empires, the foot measure has been every where retained, but infinitely varied in its extent. Every nation of modern Europe has a foot measure, no two of which are the same. The English foot indeed was adopted and established in Russia by Peter the Great; but the original Russian foot was not the same. The Hebrew shekel and maneh, the Greek mina, and the Roman pondo, were *weights*. The general name *weight* improperly applied to the specific unit of weight. The Latin word libra, still more improperly, was borrowed from the balance in which it was employed: libra was the balance, and at the same time the pound weight. The terms *weight* and *balance* were thus generic terms, without specific meaning. They signified any weight in the balance, and varied according to the varying gravities of the specific standard unit at different times and in different countries. When, by the debasement of the coins they ceased to be identical with the weights, they still retained their names. The *pound* sterling retains its name three centuries after it has ceased to exist as a weight, and after having, as money, lost more than two thirds of its substance. We have discarded it indeed from our vocabulary; but it is still the unit of moneys of account in England. The *livre* tournois of France, after still greater degeneracy, continued until the late revolution, and has only been laid aside for the new system. The ounce, the drachm, and the grain, are specific names, indefinitely applied as indefinite parts of an indefinite whole. The English pound avoirdupois is heavier than the pound troy; but the ounce avoirdupois is lighter than the ounce troy. The weights and measures of all the old systems present the perpetual paradox of a whole not equal to all its parts. Even numbers lose the definite character which is essential to their nature. A dozen become sixteen, twenty-eight signify twenty-five, one hundred and twelve mean a hundred.

The indiscriminate application of the same generic term to different specific things, and the misapplication of one specific term to another specific thing, universally pervade all the old systems, and are the inexhaustible fountains of diversity, confusion, and fraud. In the vocabulary of the French system, there is one specific, definite, significant word, to denote the unit of lineal measure; one for superficial, and one for solid measure; one for the unit of measures of capacity, and one for the unit of weights. The word is exclusively appropriated to the thing, and the thing to the word. The metre is a definite measure of length: it is nothing else. It cannot be a measure of one length in one country, and of another length in another. The gramme is a specific weight, and the litre a vessel of specific cubic contents, containing a specific weight of water. The multiples of these units are denoted by prefixing to them syllables derived from the Greek language, significant of their increase in decimal proportions. Thus, ten metres form a deca-metre; ten grammes, a deca-gramme; ten litres, a deca litre. The subdivisions, or decimal fractions of the unit, are equally significant in their denominations, the prefixed syllables being derived from the Latin language. The deci-metre is a tenth part of a metre; the deci-gramme, the tenth part of a gramme; the deci-litre, the tenth part of a litre. Thus, in continued multiplication, the hecto-metre is a hundred, the kilo-metre a thousand, and the myria-metre ten thousand metres; while, in continued division, the centi-metre is the hundredth, and the milli-metre the thousandth part of the metre. The same prefixed syllables apply equally to the multiples and divisions of the weight, and of all the other measures. Four of the prefixes for multiplication, and three for division, are all that the system requires. These twelve words, with the *franc*, the *decime*, and the *centime*, of the coins, contain the whole system of French metrology, and a complete language of weights, measures, and money.

But where is the steam engine of moral power to stem the stubborn tide of prejudice, and the headlong current of inveterate usage? The cheerful, ready, and immediate adoption, by the mass of the nation, of these twelve words, would have secured the triumph of the new system of France. The unutterable confusions of signifying the same thing by different words, and different things by the same word, would have ceased. The *setier* would no longer have been a common representative for twelve boisseaus of corn, for fourteen of oats, for sixteen of salt, and for thirty-two of coal, and for eight pints of wine. The pound would no longer have been of ten, of twelve, of fourteen, of sixteen, and of eighteen ounces, in different parts of the same country. The weights and the measures would have been both perfect and just: and the blessing of uniformity enjoyed by France would have been the most effective recommendation of her system to all the rest of mankind. It is mortifying to the philanthropy, which yearns for the improvement of the condition of man, to know that this is precisely the part of the system which it has been found impracticable to carry through.

ON WEIGHTS AND MEASURES.

The modern language of all the mathematical and physical sciences is derived from the Greek and Latin; with a partial exception of some terms which are of Arabic origin. Geography, chemistry, the pure mathematics, botany, mineralogy, zoology, in all of which great discoveries have been made within the last three centuries, have borrowed from those primitive languages almost invariably the words by which those discoveries have been expressed. They are the languages in which all that was heretofore known of art or science was contained: nor are the moral and political sciences less indebted to them for numerous additions to their vocabalaries which the progress of modern improvements has required. But there is a natural aversion in the mass of mankind to the adoption of words, to which their lips and ears are not from their infancy accustomed. Hence it is that the use of all technical language is excluded from social conversation, and from all literary composition suited to general reading; from poetry, from oratory, from all the regions of imagination and taste in the world of the human mind. The student of science, in his cabinet, easily familiarizes to his memory, and adopts without repugnance, words indicative of new discoveries or inventions, analogous to the words in the same science already stored in his memory. The artist, at his work, finds no difficulty to receive or use the words appropriate to his own profession. But the general mass of mankind, of every condition, reluct at the use of unaccustomed sounds, and shrink especially from new words of many syllables. But weights and measures are instruments indispensable, not only to the philosophical student and the professional artist; they are the want of every individual and of every day. They are the want of food, of raiment, of shelter, of all the labors and all the pleasures of social existence. Weights and measures, like all the common necessaries of life, have, in all the countries of modern Europe, customary names of one, or, at most, of two, syllables. The units of the new French system have no more; but their multiples and subdivisions have four or five; and, although compounded of syllables familiar to those who had any acquaintance with the classical languages of Greece and Rome, they had a strange and outlandish sound to the ears of the people in general, who would never be taught to pronounce them. Hence, after an experience of several years, it was found necessary, not only to give back to the people the vulgar fractions of their measures, which had been taken from them, but all their indefinite and many-meaning words of pound and ounce, foot, aune and thumb, boisseau and pint. Since which time there have been, besides all the relics of the old metrology, two concurrent systems of weights and measures in France; one, the proper legal system, with decimal divisions and multiplications, and the new, precise, and significant nomenclature; and the other a system of sufferance, with the same instruments, but divided in all the old varieties of vulgar fractions, and with the old improper vocabulary, made still more so by its adaptation to new and different things.

Perhaps it may be found, by more protracted and multiplied experience, that this is the only uniformity attainable by a system of weights and measures for universal use : that the same material instruments shall be divisible decimally for calculations and accounts; but in any other manner suited to convenience in the shops and markets; that their appropriate legal denominations shall be used for computation, and the trivial names for actual weight, or mensuration.

It results, however, from this review of the present condition of the French system in its native country, and from the comparison of its theoretical advantages over that which we already possess, that the time has not arrived at which so great and hazardous an experiment can be recommended, as that of discarding all our established existing weights and measures, to adopt and legalize those of France in their stead. The single standard, proportional to the circumference of the earth; the singleness of the units for all the various modes of mensuration; the universal application to them of decimal arithmetic; the unbroken chain of connection between all weights, measures, moneys, and coins; and the precise, significant, short, and complete vocabulary of their denominations; altogether forming a system adapted equally to the use of all mankind; afford such a combination of the principle of uniformity for all the most important operations of the intercourse of human society; the establishment of such a system so obviously tends to that great result, the improvement of the physical, moral, and intellectual, condition of man upon earth; that there can be neither doubt nor hesitation in the opinion, that the ultimate adoption, and universal though modified application of that system, is a consummation devoutly to be wished.

To despair of human improvement is not more congenial to the judgment of sound philosophy than to the temper of brotherly kindness. Uniformity of weights and measures is, and has been for ages, the common, earnest, and anxious pursuit of France, of Great Britain, and, since their independent existence, of the United States. To the attainment of one object, common to them all, they have been proceeding by different means, and with different ultimate ends. France alone has proposed a plan suitable to the ends of all; and has invited co-operation for its construction and establishment. The associated pursuit of great objects of common interest is among the most powerful modern expedients for the improvement of man. The principle is at this time in full operation, for the abolition of the African slave-trade. What reason can be assigned, why other objects, of common interest to the whole species, should not be in like manner made the subject of common deliberation and concerted effort? To promote the intercourse of nations with each other, the uniformity of their weights and measures is among the most efficacious agencies: and this uniformity can be effected only by mutual understanding and united energy. A single and universal system can be finally established only by a general convention, to which the principal nations of the world shall be parties, and to which they shall all give their assent. To effect this, would seem to be no difficult achievement. It has one advantage over every plan of moral or political improvement, not excepting the abolition of the slave-trade itself: there neither is, nor can be, any great counteracting *interest* to overcome. The conquest to be obtained is merely over prejudices, usages, and perhaps national jealousies. The whole evil to be subdued is diversity of opinion with regard to the means of attaining the same end. To the formation of the French system, the learning and the genius of other nations did co-operate with those of her native sons. The co-operation of Great Britain was invited; and there is no doubt that of the United States would have been accepted, had it been offered. The French system embraces all the great and important principles of uniformity, which can be applied to weights and measures: but that system is not yet complete. It is susceptible of many modifications and improvements. Considered merely as a labor-saving machine, it is a new power, offered to man, incomparably greater than that which he has acquired by the new agency which he has given to steam. It is in design the greatest *invention* of human ingenuity since that of printing. But, like that, and every other useful and complicated invention, it could not be struck out perfect at a heat. Time and experience have already dictated many improvements of its mechanism; and others may, and undoubtedly will, be found necessary for it hereafter. But all the radical principles of uniformity are in the machine: and the more universally it shall be adopted, the more certain will it be of attaining all the perfection which is within the reach of human power.

Another motive, which would seem to facilitate this concert of nations, is, that it conceals no lurking danger to the independence of any of them. It needs no convocation of sovereigns, armed with military power. It opens no avenue to partial combinations and intrigues. It can mask, under the vizor of virtue, no project of avarice or ambition. It can disguise no private or perverted ends, under the varnish of generous and benevolent aims. It has no final appeal to physical force; no *ultima ratio* of cannon balls. Its objects are not only pacific in their nature, but can be pursued by no other than peaceable means. Would it not be strange, if, while mankind find it so easy to attain uniformity in the use of every engine adapted to their mutual destruction, they should find it impracticable to agree upon the few and simple but indispensable instruments of all their intercourse of peace and friendship and beneficence—that they should use the same artillery and musketry, and bayonets and swords and lances, for the wholesale trade of human slaughter, and that they should refuse to weigh by the same pound, to measure by the same rule, to drink from the same cup, to use in fine the same materials for ministering to the wants and contributing to the enjoyments of one another?

These views are presented as leading to the conclusion, that, as final and universal uniformity of weights and measures is the common desideratum for all civilized nations; as France has formed, and for her own use has established, a system, adapted, by the highest efforts of human science, ingenuity, and skill, to the common purposes of all; as this system is yet new, imperfect, susceptible of great improvements, and struggling for existence even in the country which gave it birth; as its universal establishment would be a universal blessing; and as, if ever effected, it can only be by consent, and not by force, in which the energies of opinion must precede those of legislation; it would be worthy of the dignity of the Congress of the United States to consult the opinions of all the civilized nations with whom they have a friendly intercourse; to ascertain, with the utmost attainable accuracy, the existing state of their respective weights and measures; to take up and pursue, with steady, persevering, but always temperate and discreet exertions, the idea conceived, and thus far executed, by France, and to co-operate with her to the final and universal establishment of her system.

But, although it is respectfully proposed that Congress should immediately sanction this consultation, and that it should commence, in the first instance, with Great Britain and France, it is not expected that it will be attended with immediate success. Ardent as the pursuit of uniformity has been for ages in England, the idea of extending it beyond the British dominions has hitherto received but little countenance there. The operation of changes of opinion there is slow; the aversion to all innovations, deep. More than two hundred years had elapsed from the Gregorian reformation of the calendar, before it was adopted in England. It is to this day still rejected throughout the Russian empire. It is not even intended to propose the adoption by ourselves of the French metrology for the present. The reasons have been given for believing, that the time is not yet matured for this reformation. Much less is it supposed adviseable to propose its adoption to any other nation. But, in consulting them, it will be proper to let them understand, that the design and motive of opening the communication is, to promote the final establishment of a system of weights and measures, to be common to all civilized nations.

In contemplating so great, but so beneficial a change, as the ultimate object of the proposal now submitted to the consideration of Congress, it is supposed to be most congenial to the end, to attempt no present change whatever in our existing weights and measures; to let the standards remain precisely as they are; and to confine the proceedings of Congress at this time to authorizing the Executive to open these communications with the European nations where we have accredited ministers and agents, and to such declaratory enactments and regulations as may secure a more perfect uniformity in the weights and measures now in use throughout the Union.

The motives for entertaining the opinion, that any change in our system at the present time would be inexpedient, are four:

First, That no change whatever of the system could be adopted, without losing the greatest of all the elements of uniformity, that referring to the persons using the same system. This uniformity we now possess, in common with the whole British nation; the nation with which, of all the nations of the earth, we have the most of that intercourse which requires the constant use of weights and measures. No change is believed possible, other than that of the whole system, the benefit of which would compensate for the loss of this uniformity.

Secondly, That the system, as it exists, has an uniformity of proportion very convenient and useful, which any alteration of it would disturb, and perhaps destroy; the proportion between the avoirdupois and troy weights, and that between the avoirdupois weight and the foot measure; one cubic foot containing of spring water exactly one thousand ounces avoirdupois, and one pound avoirdupois consisting of exactly seven thousand grains troy.

Thirdly, That the experience of France has proved, that binary, ternary, duodecimal, and sexagesimal divisions, are as necessary to the practical use of weights and measures, as the decimal divisions are convenient for calculations resulting from them; and that no plan for introducing the latter can dispense with the continued use of the former.

Fourthly, That the only *material* improvement, of which the present system is believed to be susceptible, would be the restoration of identity between weights and silver coins; a change, the advantages of which would be very great, but which could not be effected without a corresponding and almost total change in our coinage and moneys of account; a change the more exceptionable, as our monetary system is itself a new, and has hitherto been a successful institution.

Of all the nations of European origin, ours is that which least requires any change in the system of their weights and measures. With the exception of Louisiana, the established system is, and always has been, throughout the Union, the same. Under the feudal system of Europe, combined with the hierarchy of the church of Rome, the people were in servitude, and every chieftain of a village, or owner of a castle, possessed or asserted the attributes of sovereign power. Among the rest, the feudal lords were in the practice of coining money, and fixing their own weights and measures. This is the great source of numberless diversities existing in every part of Europe, proceeding not from the varieties which in a course of ages befell the same system, but from those of diversity of origin. The nations of Europe are, in their origin, all compositions of victorious and vanquished people. Their institutions are compositions of military power and religious opinions. Their doctrines are, that freedom is the grant of the sovereign to the people, and that the sovereign is amenable only to God. These doctrines are not congenial to nations originating in colonial establishments. Colonies carry with them the general laws, opinions, and usages, of the nation from which they emanate, and the prejudices and passions of the age of their emigration. The North American colonies had nothing military in their origin. The first English colonies on this continent were speculations of commerce: they commenced precisely at the period of that struggle in England between liberty and power, which, after long and bloody civil wars, terminated in a compromise between the two conflicting principles. The colonies were founded by that portion of the people, who were arrayed on the side of liberty. They brought with them all the rights, but none of the servitudes, of the parent country. Their constitutions were, indeed, conformably to the spirit of the feudal policy, charters granted by the crown; but they were all adherents to the doctrine, that charters were not donations, but compacts. They brought with them the weights and measures of the law, and not those of any particular district or franchise. The only change which has taken place in England with regard to the legal standards of weights and measures, since the first settlement of the North American colonies, has been the specification of the contents of measures of capacity, by prescribing their dimensions in cubical inches. All the standards at the exchequer are the same that they were at the first settlement of Jamestown; with the exception of the wine gallon, which is of the time of queen Anne: and the standards of the exchequer are the prototypes from which all the weights and measures of the Union are derived.

A particular statement of the regulations of the several states relative to weights and measures, is subjoined to this report, in the appendix.

The first settlement of the English colonies on the continent of North America was undertaken towards the close of the reign of queen Elizabeth, in honor of whom it received the name of Virginia.

During the same reign of Elizabeth, and cotemporaneous with the adventures which preceded the settlement of Jamestown, the act of parliament of 1592 passed, defining in feet the statute mile. This mile, together with its elementary units, the foot and inch, were the measures by which all the territories, granted by the successors of Elizabeth, in this hemisphere, were defined. The foot and inch, from usage immemorial in England, and by a statute then of more than three centuries' antiquity, had been the elements of superficial, as well as of itinerary land measure. These, therefore, were not only the most natural measures for the use of the English colonies; they were inwoven in their primitive constitutions, and were brought with their charters, an essential part of their possessions.

Among the earliest traces of colonial legislation in Virginia and in New England, we find acts declaring the assize of London, and the standards of the exchequer, to be the only lawful prototypes of the weights and measures of the colonies. The foot and inch were of dimensions perfectly well ascertained: and in the year 1601, only seven years before the settlement at Jamestown, and less than twenty before that of Plymouth, new standards, not only of the yard and ell, but of the avoirdupois and troy weights, and of the bushel, corn gallon, quart, and pint, had been deposited at the exchequer. There was neither uncertainty, nor perceptible diversity, with regard to the long measures or the weights; but the standard vessels of capacity were of various dimensions. The bushel of 1601 contained 2,124 cubic inches: it was therefore a copy from an older standard, made in exact conformity to the rule prescribed in the statute of 1266, and very probably the identical standard therein described. It contained eight corn gallons of wheat, equiponderant to eight Irish gallons of Gascoign wine; of wheat, thirty-two kernels of which were of equal weight with the round, unclipped penny sterling of 1266. Its corresponding wine gallon, therefore, would have been the Irish gallon of 217.6 cubic inches; and its corresponding corn gallon of 265.5 inches, an intermediate between the Rumford quart and gallon of 1228, and differing less than one inch from either of them. There were two other standard bushels at the exchequer, of the same dimensions; one of the age of Henry the Seventh, and one dated 1091. This has been supposed to be a mistake for 1591 or 1601. But as it is not probable that two standard bushels should have been deposited in the exchequer at the same time, or even at dates so near to each other, a conjecture may be indulged, that the 1091 marks the date, when the standard measure, described in the statute of 1266 was made. Of that standard, these three bushels were unquestionably copies.

The corn and the ale gallons of 1601 were of 272 cubic inches; and there was one of Henry the Seventh there, of the same size, as reported by the artist who measured them for the commissioners of the excise in 1688. When measured again by order of the committee of the House of Commons, in 1758, they were reported to contain each about one inch less. The true size intended for all of them was 272; and they were made by an application of the rule of 1266 to the troy weight wheat of the act of 1496. They were the eighth parts of a bushel of 2,176 inches; and their corresponding wine gallon was the Guildhall gallon of 224 inches.

There were, in 1601, a standard quart of 70 inches, and a pint of 34.8; which were evidently intended to be in exact proportions to each other: and the gallon, to which they referred, was the gallon of 282 inches. This would have made a bushel of 2,256 inches; and its corresponding wine gallon is of 231 inches. The standards, thus made, were by an application both of the wheat and of the rule described in the statute of 1266 to the troy weight gallon of 1496; that is, the wheat was of the kind, 32 kernels of which weighed the same as the old penny sterling, and of which the wine gallon contained eight pounds troy weight. There was a standard bushel of Henry the Seventh at the exchequer, of 2,224 inches, probably the bushel from which this quart and these pints were deduced.

There was also the *Winchester* bushel of 2,145.6 cubic inches, made in the reign of Henry the Seventh, but from its name evidently copied from a standard which had been kept at Winchester when that place was the capital of the kingdom. This bushel had been made, by combining the rule of 1266 with the assize of casks which, in the statute of 1423, is declared to be of *old time*, by which the hogshead, or eight cubic feet of Gascoign wine, consisted of 63 gallons. That hogshead was a quarter of a ton of wine, as eight Winchester bushels contained a quarter of a ton of wheat. The gallon was of $219\frac{1}{2}$ cubic inches; and the corresponding ale gallon was of 268.2 inches. There was at the exchequer no wine or ale gallon of those dimensions; because the wine gallon of 224 inches, and the corn gallon of 272, made under the statutes of 1496 and 1531, had been substituted in their stead. At the exchequer, there was indeed no wine gallon at all. Those of older date than the act of 1496 had disappeared, and the gallon of 224 inches made according to that act, had been delivered out of the exchequer to the city of London, and was at Guildhall.

Such was the state of the standards in London, at the time of the first colonial emigrations to this continent.

MASSACHUSETTS.

Among the colony laws of Massachusetts, there is an act of the year 1647, directing the country treasurer *to provide,* at the country's charge, weights and measures *of all sorts* for continual standards. In the specification which ensues in the act, all the measures, of which there were standards at the exchequer, are mentioned, with special discrimination of *wine* and *ale measures;* but the weights only *after sixteen ounces* to the pound, are named. They then had no occasion for the troy weights.

At a still earlier date, in 1641, it had been prescribed that all casks for any liquor, fish, beef, pork, or other commodities to be put to sale, should be *of London assize:* and in 1646 a corresponding assize of *staves* had been ordained.

The law of 1647 did not expressly direct where the treasurer was to procure the standards: but the Exchequer and Guildhall were the only places where they were to be obtained; and, from subsequent acts, the fact appears that they were obtained there.

At the first session of the general court under the charter of William and Mary, in 1692, two laws were enacted; one, re-ordaining the London assize of casks, and specifying that the butt should contain 126 gallons, the puncheon 84, the hogshead 63, the tierce 42, and the barrel 31½ gallons; the other, for due regulation of weights and measures, declaring that the brass and copper measures, *formerly sent* out of England, with certificate out of the exchequer to be approved *Winchester* measure, according to the standard in the exchequer, should be the public allowed standard throughout the province for the proving and sealing all weights and measures thereby, and re-enacting, with an additional clause, the colonial law of 1647.

An act of the year 1700 prescribes, that the bushel used for the sale of meal, fruits, and other things, usually sold by heap, shall be not less than 18½ inches wide within side; the half bushel not less than 13¾ inches; the peck not less than 10¾, and the half peck not less than 9 inches.

It is very remarkable that this law was enacted one year *before* the act of parliament of 13 William III. which gives and prescribes in cubical inches the dimensions of the Winchester bushel. The object of the provincial law was, to prohibit the use of bushels, which, though of the same cubical capacity, should be of shorter diameter and greater depth. It was for the benefit of the heap. It prescribed, therefore, only the diameter, without mentioning the depth; but that diameter, for the bushel, is identically the same, 18½ inches, as the act of parliament of the ensuing year declares to be the width of the Winchester bushel in the exchequer. As the provincial standard must have been the model from which the law of the province took its measure of a diameter, its perfect coincidence with the subsequent definition of the act of parliament, is a proof of the correctness of the copy from the Winchester bushel of the exchequer.

In 1705, the treasurer of the province was required by law to procure a beam, scale, and a nest of *troy weights* from 128 ounces down, marked with a mark or stamp used at the exchequer, for a public standard. Every town was to be provided with a nest of troy weights of different form from the avoirdupois: and a penalty was annexed to the use of any other than sealed troy weights, for weighing silver, bullion, or other species whatsoever, proper and *used* to be weighed by troy weights.

In the year 1707, there was an act of parliament, 6 Anne, ch. 30, " for ascertaining the rates of foreign coins in her majesty's planta-" tions in America." It had been preceded, in 1704, by a proclamation of the queen, declaring the value of many foreign silver coins, and particularly of the Spanish piece of eight, or dollar. At that period, as in a certain degree at the present, the Spanish dollar and its parts formed the principal circulating coins of this country. The act declares the value of the Seville, Pillar, and Mexican pieces of eight, to be four shillings and six-pence sterling, and their weight to be seventeen pennyweights and a half, or 420 grains. It forbids their being taken in the colonies at more than six shillings each: and this act constituted what, from that time till the period of the Revolution, in Virginia and New England, was denominated " lawful " money." The act itself was published in the province of Massachusetts Bay with the statutes of the provincial legislature, as was practised with regard to all the acts of parliament, the authority of which was recognized. It may be here incidentally remarked, that the laws of Congress, which estimate the value of the English pound sterling at four dollars and forty-four cents, are all founded upon the proportions established by this act, although the weight and value, both of the dollar and of the shilling and pound sterling, have since that time been changed. Some further observations on this subject are submitted in the appendix: from which it will appear, that the real value of our silver dollar, in the silver English half crowns or shillings of this time, is four shillings, seven pence, and nearly one farthing; and that the pound sterling of such actual English silver coins is, in the silver money of the United States, not 4 dollars 44 cents, but only 4 dollars, 34 cents, and 9 mills.

In the year 1715 a light house was built at the entrance of Boston harbor; and a tonnage duty being levied upon vessels entering the harbor, to defray the expense of building and supporting it, the rule of measurement for ascertaining the tonnage, prescribed by the act, was, that a vessel of two decks should be measured upon the main deck, from the stem to the stern post, then subducting the breadth, from outside to outside, athwart the main beam, the remainder to be accounted her length by the keel, which, being multiplied by the breadth, and the product by one half the breadth for the depth, and the whole product divided by 100, the quotient was to be accounted the tonnage of the ship. Vessels of a single deck, or 1½ deck, were to be measured in the same manner, except the depth in hold, which was to be from the under side of the main beam to the ceiling.

In 1730 a new set of brass and copper avoirdupois weights, and of measures, was imported from the Exchequer, with certificate of their being approved Winchester measure, according to the standard in the Exchequer. These were, by a new statute, declared to be the public standards of the province: and they continue to be those of the commonwealth at this day. It does not appear that the troy weights were renewed at the same time. The standards of them had been imported only twenty-five years before, and could not need renewing.

In 1751 the act of parliament introducing the Gregorian calendar was adopted, in the usual manner, by inserting it among the laws of the provincial legislature.

Since the Revolution all the laws of the province have been revised: and, by an act of the Legislature of 26th February, 1800, all the principal regulations concerning weights and measures were renewed and confirmed.

This law declares that the brass and copper measures, *formerly* (1730) sent out of England with a certificate from the Exchequer, shall be and remain the public standards throughout the commonwealth: and it requires the treasurer of the commonwealth to cause to be had and preserved a complete set of new beams, weights, avoirdupois and troy, and measures of length and of capacity, wet and dry, to be used only as public standards. This act is to continue until Congress shall have fixed by law the standard of weights and measures.

A statute of 9th March, 1804, recites, that the troy weights used by the treasurer of the commonwealth, as state standards, had, by long use, diminished and undergone an alteration in their proportions. (They had then been just one century in use.) It directs him, therefore, to add, or cause to be added, a specified number of grains to each of the weights, from that of 128 ounces to the half ounce; or to procure new weights of the same denomination, and conformable to the state standards, with such additions: which weights, so corrected, are declared to be the standards of troy weight for the commonwealth. By information from various sources, it is known that the standards of the state of Massachusetts are, at this time, perfectly conformable to those of the Exchequer.

There are a multitude of laws regulating the assize of casks, assigning different dimensions for containing different articles. They generally prescribe the length of the staves within the chime, and sometimes the diameter of the heads. They also specify the weight of the article which the cask is to contain. Staves are an article of exportation; and their length, breadth, and thickness, are regulated by law.

NEW HAMPSHIRE AND VERMONT.

The laws of New Hampshire, and of Vermont, relating to weights and measures, appear to have been modelled upon those of Massachusetts. In both these states the standards are required to be according to the approved Winchester measures, allowed in England, in the Exchequer. The first act of New Hampshire to that effect was of 13th May, 1718, and the last of 15th December, 1797. The statute of Vermont is of the 8th of March, 1797. Neither New Hampshire nor Vermont has established the authority of the troy weights by law.

RHODE ISLAND.

Rhode Island has no statute upon the subject. Her weights and measures are, however, the same, and her standards are taken from those of Massachusetts.

CONNECTICUT.

In the laws and standards of Connecticut there are peculiarities deserving of remark.

A statute of October, 1800, contains the following provisions: "That the brass measures, the property of this state, kept at the "Treasury, that is to say, a half bushel measure, containing one "thousand and ninety-nine cubic inches, very near, a peck measure "and half peck measure, when reduced to a just proportion, be the "standard of the corn measures in this state, which are called by "those names respectively; that the brass vessels ordered to be pro-"vided by this Assembly [one of the capacity of two hundred and "twenty-four cubic inches]* and the other of the capacity of two "hundred and eighty two cubic inches, shall be, when procured, the "first of them the standard of a wine gallon, and the other the stand-"ard of an ale or beer gallon, in this state; that the iron, or brass "rod or plate, ordered by this Assembly to be provided, of one yard "in length, to be divided into three equal parts, for feet, in length, "and one of those parts to be subdivided into twelve equal parts, for "inches, shall be the standard of those measures respectively; and "that the brass weights, the property of the state, kept at the Trea-"sury, of one, two, four, seven, fourteen, twenty-eight and fifty-six "pounds, shall be the standard of avoirdupois weight in this state."

---

* These brackets, in the printed volume of the laws of Connecticut, indicate that the part enclosed has been repealed.

A subsequent section (5) requires of the selectmen of each town to provide town-standards, of good and sufficient materials, which, for the standards of liquid measure, shall be copper, brass, or pewter; also, vessels for corn measure, of forms and dimensions thus described: "A two quart measure, the bottom of which, on the inside, is "four inches wide on two opposite sides, and four inches and a half "on the two other sides, and its height from thence seven inches and "sixty-three hundredths of an inch;" [137.34 cubic inches.] A quart measure of "three inches square from bottom to top, through- "out, and its height seven inches and sixty-three hundredths of an "inch;" [68.67 cubic inches.] A pint measure of three inches square from bottom to top throughout, and its height three inches and eighty-two hundredths of an inch; [34.38 cubic inches.]

The assumption of the old Guildhall wine gallon, of 224 inches, in this act, is the more surprising, inasmuch as a colonial statute of the year 1752 had already established the gallon of 231 inches. What the occasion of it was, has not been ascertained; but it was probably taken from an existing standard, which had been originally taken from the Guildhall gallon. Whatever the cause of it may have been, this part of the act was repealed the next year, (October, 1801) and the Treasurer was directed, without delay, to provide a vessel of brass, of five inches square from bottom to top throughout, and nine inches and twenty-four hundredths of an inch in height, containing two hundred and thirty-one cubic inches, which was declared the standard wine gallon of the state.

The half bushel measure, which in 1800 was the property of the state, kept at the treasury, containing 1099 cubic inches very near, was of course not originally derived from the Winchester bushel. By the colonial laws of Connecticut, it appears, that, as early as the year 1670, there were *colony standards* kept at Hartford: and the half bushel, which in the year 1800 was there at the treasury, the property of the state, was either one of those same standards of 1670, or a copy from it. That it was not borrowed from the Massachusetts standards is also manifest, because the Massachusetts bushel was copied from the Winchester bushel. It may be concluded, with great probability, that the Connecticut half bushel was first taken from the bushel in the exchequer of Henry the Seventh with a copper rim; though it contains thirteen cubic inches less than in proportion to that standard. This difference, in so large a measure, may have been the effect of very slight inaccuracy in the first copy, increased by the decay or the change of the vessel. The bushel with the copper rim was deposited at the exchequer after the act of 1496; and was made from the wine gallon of that act, with the rule of the act of 1266, and the pound of fifteen ounces troy weight. The quart and pint at the exchequer of 1601, were formed from this bushel. The pint differs less than half an inch from that prescribed by this act of Connecticut of 1800.

In the laws of Connecticut, as in those of New Hampshire and Vermont, there is no formal establishment or recognition of troy weights; nor does there appear to be any standard of them existing in the state. But in the lists of rateable estate, prescribed by the laws of Connecticut, silver plate is estimated at one dollar eleven cents per ounce, which must obviously be intended the ounce troy.

The assize of casks is regulated by various laws: and the dimensions of the barrel for packing salted provisions for exportation are the same as those established in Massachusetts and New York. The London assize of *tight* casks, from the puncheon of 126 to the barrel of 31½ gallons, was co-eval with the first legislation of the colony; and was re-enacted by a statute of 1795. It expressly declares that these gallons shall be of 231 cubic inches; and directs that they shall be computed by taking, in inches and decimal parts of an inch, the bulge or bung diameter, each head diameter, and the length within the cask, with Gunter's rule of gauging.

The assize of staves is the same as in Massachusetts.

By an act of October 1796, the standard weight of wheat is declared to be sixty pounds nett to the bushel.

NEW YORK.

New York was originally the seat of a colony from the Netherlands, the settlers of which doubtless brought with them the weights and measures of their own country. Towards the close of the seventeenth century it fell into the possession of the English; and on the 19th of June, 1703, an act of the colonial legislature established all the English weights and measures, *according to the standards in the exchequer*. This act was drawn with great care, and evidently with the purpose of embracing all the provisions of the then existing English statutes, regulating weights, measures, and casks, particularly those of 1266, 1304, 1439, and 1496, without being aware of the utter incompatibility of those statutes with one another.

Instead, however, of adopting in terms the London assize of casks, from the ton of 252 gallons downwards, this act prescribes in inches the length and head diameters of the various casks; and, by a very remarkable peculiarity, changes the names of all the dry casks. It directs that

The Hogshead shall be 40 inches long,	33 inches in the bulge,	27 inches in the head.
Tierce . . 36	. . 27	 23
Barrel . . 30	. . 26	 22
Half barrel 25	. . 20	. . . 16
Quarter barrel . 20	. . 16	 13

But it adds, that *the* barrels shall contain 31½ gallons wine measure, or within half a gallon more or less, and all other casks in proportion. This last provision adopted the whole London assize for tight casks. But the dimensions prescribed for the *hogshead*, give a cask of about 126 gallons, which, in the London assize, made the butt or pipe: and thus the New York tierce was of 80 gallons, which constituted the *real* contents of the London puncheon; the New York barrel was of 60 gallons, answering to the London hogshead; and the New York half barrel of 30 gallons, to the London barrel

On the 10th of April, 1784, the legislature of New York passed an act to *ascertain* weights and measures within the state. It declares the standard weights and measures which were in the custody of William Hardenbrook, public sealer and marker in the city and county of New York, at the time of the declaration of Independence, which were according to the standard of the exchequer, to be the standards throughout the state. William Hardenbrook was directed to deliver them to the clerk of the city and county of New York, and to make oath that they were the same which he had received from the court of exchequer.

By an act of 7th March, 1788, the standard weight of wheat brought to the city of New York for sale, was fixed at sixty pounds nett to the bushel.

On the 24th of March, 1809, passed an act relative to a standard of long measure, and for other purposes. It declares a brass yard measure, engraved and sealed at the exchequer of Great Britain, procured in 1803 by the corporation of New York, presented to the state, and deposited, with authenticating documents, in the secretary's office, to be the standard yard measure of the state.

The last statute, upon this subject, of New York, is an act, to regulate weights and measures, and passed on the 19th of March 1813; which declares that there shall be one just beam, one certain weight, and measure for distance and capacity; that is to say, avoirdupois and troy weights, bushels, half-bushels, pecks, half-pecks, and quarts; and gallons, half-gallons, quarts, pints, and gills; and one certain rod tor long measure, according to " the standard in use in the state " on the day of the declaration of the Independence thereof, and that " the standard of weights and measures in the office of the Secretary of " the State, which is according to the standard in the court of ex- " chequer in that part of Great Britain called England, shall be and " remain the standard for ascertaining all beams, weights and mea- " sures throughout the state, until the Congress of the United States " shall establish the standard of weights and measures for the United " States."

The assize of casks continues as it was regulated by the act of 1703: but a variety of special statutes assign dimensions different from it for barrels in which beef, pork, fish, flour, pot and pearl ashes, &c. are packed for exportation. These, as in the New England states, are adapted to contain a certain specified weight of each article. The assize of *staves* regulated by an act of 26th March, 1813, is substantially the same as that of Massachusetts: and as the capacity of the barrel must always depend in a great degree upon the size of the staves and heading of which it is made, the contents of all these barrels vary little from 30 gallons wine measure of 231 cubic inches equal to 6,930 inches.

NEW JERSEY.

In New-Jersey, which was originally a part of the Dutch settlement, the English weights and measures were established at a later period than in New York. An act of the colonial legislature, of 13th August, 1725, recites, in its preamble, that nothing is more agreeable to common justice and equity than that throughout the province there should be *one* just weight and balance, one true and perfect standard for measures, for *want* whereof experience had shown that many frauds and deceits had happened; for remedy of which, it establishes, in the first section, an assize of casks for packing of beef and pork, since altered; and in the second, declares, that there shall be one just beam and balance, one certain standard for " weights, that " is to say : for avoirdupois and troy weights, one standard for mea- " sures, bushels, half bushels, pecks, and half-pecks, one just stand- " ard for liquid measures, that is to say : wine and beer measure ; " and one yard ; all which shall be according to the standard of the " exchequer in Great Britain."

The phraseology of this statute has some resemblance to that of the 25th chapter of Magna Charta, and may serve as a lucid commentary upon it; for, although its avowed object is uniformity, and even *unity* of standard, it expressly sanctions two weights, avoirdupois and troy, and two liquid measures for wine and beer. This statute also, as well as that of New York of 1813, shows that the term *gallon* is improperly used when applied to dry measure, its real denomination being that of half-peck.

The laws of New Jersey relating to the assize of barrels have been various. By an act of 1774, revived in 1783, the barrel is required to contain 31½ wine gallons, and not ½ a gallon more or less ; half-barrels 16 gallons, and not one quart more or less. The assize of staves (26th September, 1772,) is materially the same as in all the eastern states.

PENNSYLVANIA.

In the year 1700, two laws relating to weights and measures were enacted by the colonial legislature. The first [laws of Pennsylvania, Bioren's edition, vol. 1, page 18] ordains, that brass standards of weights and measures, according to the standards for the exchequer, should be *obtained*, and kept in each county. Sec. 2. That a brass half-bushel, then in Philadelphia, and a bushel and peck proportionable, and all lesser measures and weights coming from England, being duly sealed in London, or other measures agreeable therewith, should be accounted good till the standard should be obtained.. Sec. 3. That no person should sell beer or ale by retail, *but by beer measure*, according to the standard of England.

The second, not only adopted the London assize of casks, but required that *all* tight casks, for beer, ale, cider, pork, beef, and oil, and all such commodities, should be made of good, sound, well seasoned, white oak timber, and contain,

The Puncheon	84 gallons
Hogshead	63 ,,
Tierce	42 ,,
Barrel	31½ ,,
Half barrel -	16 ,,

wine measure, according to the practice of the neighboring colonies.

This act regulated the assize of staves for hogsheads and barrels; and prescribed that tobacco hogsheads should be four feet long, or within an inch more or less, 32 inches in the head, equal to the gauge of Maryland, and be four hogsheads to a ton; that the flour cask should be not above *double* the gauge of wine measure; the half barrel to be of 31½ gallons, and the barrel of 63 gallons wine measure.

As the gauge of Maryland was adopted for tobacco, so that of New York was assumed for flour, by constituting the barrel and half-barrel at double the gauge of wine measure. The origin of this must have been in the measures of the Dutch colonies, which had reference to the *last*, or double *ton* of shipping, the customary measure of the Netherlands instead of the ton.

But the most remarkable peculiarity of these two laws of Pennsylvania, enacted at the same session of the legislature, was, that while one of them applied the London assize of wine measure to the casks which were to contain *beer*, ale, and cider, the other expressly prohibited the retailers of beer and ale from selling those liquors otherwise than by beer measure; so that the retailers were obliged to buy by the small and to sell by the large measure. This inconsistency between the two statutes will not surprise us when we recollect that it occurred precisely at the time when the trial in the court of exchequer of England was litigated, concerning the duties to be paid on Mr. Thomas Barker's importation of Alicant wine. For while he, upon a claim to pay duties upon wine only by beer measure, was reducing the Attorney general, after a trial of five hours, to withdraw a juror, and cast the remedy upon parliament, the legislature of Pennsylvania, by the same erroneous application of the same name to different things, were, certainly without intention, but, in effect, enjoining upon all the publicans of the province to pay for beer by wine measure. It was a whimsical operation of the same incongruity happening in the two hemispheres at the same time, that, while Barker was struggling successfully against the supreme authority of the mother country, to pay for wine by beer measure, the Pennsylvania publicans, by the acts of their provincial legislature, were compelled to pay for beer by wine measure, and yet to be paid for it by its own.

The remedy to these disorders was applied in England and in Pennsylvania also about the same time. In both cases, however, it was partial; applied only to the special inconvenience without reaching the source of the evil. Parliament only defined the capacity of the wine gallon, fixing it at 231 cubic inches. The Pennsylvania legislature, by an act of 1705, [P. L. Bioren's edition, vol. 1, ch. 138,

p. 43,] reciting the inconsistent provisions of their two acts of 1700, and ingenuously remarking, that, in consequence of them, *retailers are obliged to sell by far greater measure than they buy*, released them from this burthensome obligation, by authorizing innkeepers to sell beer by wine measure in their houses, and by beer measure to persons to carry it out of the house. The real evil, in both cases, had proceeded from calling the two different measures of liquids by the same name. If the beer gallon had been called a *half peck*, no such questions, and no such clashing legislation, would ever have arisen. The statute of 1700, which had prescribed the London assize of casks, was repealed only in March, 1810.

The assize of staves and heading was fixed, in Pennsylvania, by a statute of 1759, [chap. 439, vol. 1, p. 222.] It was, with slight variations, the same as in all the states eastward of it. The necessary width of all staves, for exportation, was, by this act, fixed at $3\frac{1}{2}$ inches. By a subsequent act [30th March, 1803, ch. 2362, vol. 4, p. 83] staves of three inches wide are allowed as merchantable. Uninspected staves or heading may, by an act of 1790, [ch. 1501, vol. 2, p. 529] be used within the state. A great multitude of statutes in Pennsylvania, as in all the other navigating states, have regulated the assize of casks, adapting them to contain weight of the respective articles to be exported in them, and to the convenience of stowage in ships. This, as has been shewn, was the original foundation of the London assize of the ton, and of the whole English system of weights and measures: and this, in the act of Pennsylvania, of 12th September, 1789, [ch. 1422, vol. 2, p. 490,] is expressly assigned as one of the reasons for requiring casks of given dimensions.

DELAWARE.

In 1705, "An act for regulating weights and measures," directs, that each county should obtain standard brass weights and measures, according *to the queen's standards* for the exchequer; that a standard brass half-bushel should be taken from that in Philadelphia, to which the bushel and peck should be proportionable. It authorizes the use of measures and weights coming from England, duly stamped in London, or others agreeable therewith, till the standards should be procured: and it prescribes that beer or ale should be sold in retail, only by beer measure.

Subsequent acts of the legislature of Delaware define the cord of fire-wood, rate gold and silver coins by their weight in troy pennyweights and grains; and regulate the assize of casks for flour, corn, and Indian meal, in exact conformity to that of Pennsylvania.

MARYLAND.

The first act concerning weights and measures to be found in the printed editions of the statutes of this state, is of the year 1715, ch. 10. "An act relating to the standard of English weights and mea-

"sures," the preamble of which recites, that the standards are very much impaired in several of the counties of the province, and in some wholly lost or unfit for use. It therefore directs the justices of the several county courts to cause the standards they already had to be made complete, and to purchase new standards where they had none; and requires them to take security from the standard-keepers for the due execution of their office, and the safe-keeping of the standards in future.

What these standards were, is ascertained by recurrence to the records of the state for the laws, the titles only of which are given in the printed compilations of the statutes.

In 1637, at the first general assembly of which any record is extant, *a bill for corn measures* is one of forty-two which were prepared and propounded to the lord proprietary for his assent; but which were not enacted into laws, nor is there any copy of them to be found upon the record.

The next year, 1638, an act for measures and weights was one of thirty-six bills twice read and engrossed; but never read a third time, nor passed the House. There were in this bill several remarkable peculiarities. It provided that there should be one standard measure throughout the province, to be appointed by the lieutenant general, and a sealer of measures; that all contracts made for the payment of corn should be understood of corn shelled; that a barrel of new corn, tendered in payment at, or afore, the 15th of October, in any year, should be twice shaked in the barrel, and afterwards heaped as long as it will lye on; and at, or before, the feast of the nativity, should be twice shaked and filled to the edge of the barrel, or else not shaked, and heaped as before; and after the said feast it should not be shaken at all, but delivered by strike. No steelyards or other weights not sealed by the lieutenant general, or by the sealer appointed by him, were to be used, except it be small weights *sealed in England*. The act was to continue till the end of the next general assembly.

In 1641 there passed *an act for measures*, which, after reciting the inconveniences from the want of a set and appointed measure, whereby corn and other grain might be bought and sold within the province, provides, that from thenceforth *the measure used in England called the Winchester bushel* should be only used as the rule to measure all things sold by the bushel or barrel; and the barrel was to contain five such bushels. The sheriff of each county was to procure and keep such a standard bushel, whereby others should be sized and sealed, and penalties were affixed to the use of any others.

This act was to continue only two years, and then expired; but *the Winchester bushel* has, from the time of its enactment, remained the standard dry measure of Maryland.

In 1671 passed an act for providing a standard, with English weights and measures, in the several and respective counties within this province. And this statute, though omitted in all the late printed editions of the laws of Maryland, established the standard, recognized by the existing act of 1715, and by all the subsequent laws of Maryland relating to the subject.

The preamble complains, that much fraud and deceit is practised in the province, by false weights and measures: for prevention of which it enacts—

That no inhabitant, or *trader hither*, shall use in trading any other *weights* or measures than are used and made, *according to the statute of Henry the Seventh King of England* in that case made and provided, [the statute of 1496.]

That, for the discovery of abuses, nine persons, who are indicated by name, one for each county then in the province, should set up a standard at their own houses, and provide by the next shipping, or the shipping then next following at farthest, twelve half hundred weights, a quartern, half-quartern, seven pounds, four pounds, two pounds, and one pound; also, each person six stamps for making stillyards and weights, to be lettered from A to I, one letter for each county; also, each person to have nine irons, numbered from one to nine, and another with cypher, for the numbering of stillyards and pea, that they might not be changed, and to procure brass measures of ell and yard, to be sealed in England; also, a sealed bushel, half-bushel, peck, and gallon, of Winchester measure, and gallon, pottle, quart, pint, and half-pint, of wine measures, with three burnt stamps for the wooden measures and three other stamps for the pewter measures, to be all of the same letter with their other stamps; and that these weights, measures, and stamps, should be kept by those nine persons at their respective houses, to which all persons were to bring their stillyards to be tried, stamped, and numbered, once a year, and also their barrels, which were to contain five bushels, and other measures, to be sealed.

The act further provides penalties for using other weights and measures, and, in case of the death of any of the nine persons named as standard-keepers, directs that other persons should be appointed by the commissioners of the respective county courts in their stead.

The limitation of the act was to three years, or the end of the next general assembly. It was revived and continued by several successive acts till 1692, when there passed " An act for the *settling of a* " *standard* with English weights and measures within the several " and respective counties in this province."

This is in substance a re-enactment and confirmation of the statute of 1671, providing, that the justices of the county courts should, from time to time, appoint a person in each county to keep the standards, and to provide all such weights and measures as were wanting, according to the directions of the act of 1671, and an additional set for Cecil county, with stamps to be marked K.

1704, September 21, ch. 71, An act *relating* to the standard of English weights and measures, has the following preamble:

" Whereas *there is now* a standard of weights and measures *agree-*
" *able to the standard of weights and measures in her majesty's exche-*
" *quer* in England settled within the several counties of this province;"

After this preamble, the act directs, that all persons, whether inhabitants or foreigners, shall bring their stilliards, with which they weigh and receive their tobacco, every year to be tried, stamped, and numbered; and every person, trading with bushels, half bushels, &c. shall have them tried and stamped at the standard, except such as come out of England and are there stamped: and penalties are prescribed for buying or selling by stilliards or dry measures not thus tried and stamped, but they are not extended either to the weights or the liquid measures.

The titles only of all these statutes are given in the printed editions of the statutes of Maryland. But the parts of them which prescribe the standard are yet in full force. The law, is the memorable act of parliament of 1496: and the fact, in Maryland as in England, is, that the standards have been copied from those in the exchequer.

In 1765, (1st Nov. ch. 1) was passed a supplementary act to the act of 1715, already noticed, entitled " An act relating to the standard of English weights and measures."

The preamble recites, that, in the act of 1715, there is no penalty upon *buyers* by unstamped dry measures, as there is upon sellers; whence persons refuse to *buy* grain, flaxseed, and other commodities, unless by measures larger than the standard.

It, therefore, prohibits, upon £5 penalty, buying by such measures.

Neither of these two acts takes any notice either of long or liquid measures, or of weights. But the standards had been established by the statute of 1671, and have continued to this time. Beer measure appears never to have been formally established by the statute law of Maryland: but troy weight is explicitly recognized in the act of November, 1781, (ch. 16) to declare what foreign gold and silver coin shall be deemed the current money of the state. It fixes the value of several of those coins, proportionable to their weight, in ounces, pennyweights, and grains, intending, though not naming, troy weight; but rating Spanish milled pieces of eight at seven shillings and six pence, and French and English crowns at eight shillings and four pence.

In 1796, by an act to erect Baltimore, in Baltimore county, into a city, and to incorporate the inhabitants thereof, the corporation (sec. 9) are empowered to regulate and fix the assize of bread; to provide for the safe-keeping and preservation of the standard of weights and measures used within the city and precincts; also, to regulate the assize of bricks, &c. And, in 1805, by an act supplementary to the act incorporating Baltimore as a city, it is ordained, Congress not having yet fixed any standard of weights and measures, that the mayor and city council shall have and exercise the right of regulating all weights and measures within the city and precincts by the present standard, until one shall be determined on by Congress.

The assize of casks has been in Maryland, as in the other parts of the Union, both before and since our Revolution, a subject of frequent and voluminous legislation. As early as the year 1658, there had passed an act, concerning the gauge of tobacco hogsheads, which had prescribed the length and diameter at the head of those casks, the dimensions of which were then the same as those used in Virginia. In 1676, this law was re-enacted with some additional sections, and was from time to time continued until 1732.

In November, 1763, by an act for amending the staple of tobacco, &c. the hogsheads containing that article were required to be 48 inches in the length of the stave, and 70 inches in the whole diameter within the staves, at the croze and bulge; a regulation repeated in the act of November, 1801, to regulate the inspection of tobacco, which is now in force.

In 1745, there passed an act for the gauge of barrels for pork, beef, pitch, tar, turpentine, and tare of barrels for flour or bread. It did not prescribe the dimensions of flour and bread casks; but directed, that all barrels, made or used for either of those articles, should be of the size and gauge to contain at least the quantity of $31\frac{1}{2}$ gallons wine measure, and that the contents of every pork or beef barrel, for exportation or sale, should be at least 220 pounds nett of meat.

This act, though originally limited in duration to three years, and the end of the next session of the assembly, has, by successive re-enactments always limited, been continued in force to this day.

Another act, of 1786, for the inspection of salted provisions, exported and imported from and to the town of Baltimore, required that the staves of beef and pork barrels should be 29 inches long, and 18 inches diameter at the head. And these regulations, though superseded at Baltimore by the exercise of the powers vested in the corporation of that city, have been extended to other parts of the state, and are yet in force.

The size of fish barrels had been prescribed by the same act. But, in February, 1818, by an act to regulate the inspection of salted fish, it was directed, that the barrel staves should be 28 inches in length, the heads seventeen inches between the chimes, and to contain not less than 29, nor more than 31 gallons; tierces to hold not less than 45, and half-barrels not less than 15 gallons.

VIRGINIA.

Among the earliest records of the general assembly of the colony of Virginia, is an order of the 5th of March, 1623-4, that there be no weights nor measures used, but such as should be sealed by officers appointed for that purpose.

By an act of 23d February, 1631-2, it was ordained, that a barrel of corn should be accounted five bushels of *Winchester measure,* 40 gallons to the barrel. The commissioners of the monthly courts were to keep sealed barrels, and to seal such as should be brought to them. Whoever used unsealed barrels or bushels was to forfeit thirteen shillings and four pence, and sit on the pillory; and the measure and barrel deficient was to be broken and burnt. And for defective *weights*, it was ordained that the offender should be punished according to the statute in that case provided.

An act of 5th October, 1646, declares, that merchants and others, as well *Dutch* as English, practice deceit by *diversity* of weights and measures used by them; and enacts, that no merchant or trader, whether English or Dutch, shall trade with other weights and measures, than according to the statute of parliament in such cases provided. What this statute of parliament was, is explained by an act of 23d March 1661-2, which declares, that, " Whereas dayly experience sheweth that much fraud and deceit is practised in this colony by false weights and measures," for prevention thereof, no inhabitant, or trader hither, shall trade with any other weights or measures than are used and made according to the statute of 12 Henry VII. ch. 5. [the statute of 1496,] in that case provided; and that, for discovery of abuses, county commissioners shall provide sealed *weights* of half hundreds, quarternes, half quarternes, seaven pounds, fower pounds, two pounds, one pound, measures of ell and yard, of bushel, half bushel, peck, and gallon, of Winchester measure; gallon, pottle, quart, pint, half pint, of wine measure out of England; to be kept by the first of every commission at the house, and a burnt mark of (cv.) and a stamp for leaden weights and pewter potts, whither all persons, not using weights and measures brought out of England, and sealed there, shall bring all their barrels (which are to contain five bushels) and other measures to be sealed and their stillyards to be tried. Then follow penalties (in tobacco) for selling by other than sealed weights and measures, and upon commissioners for not providing standards.

Thus in Virginia, as in Maryland, the English statute of Henry VII. of 1496, has for near a century and a half been nominally the law of the land concerning weights and measures; while, at the same time, the actual weights and measures of capacity have been copies from the standards in the Exchequer, not one of which has ever been conformable to the statute of 1496. And this very act of Virginia, of 1661, while establishing by law the exclusive *troy* weight, *wine* gallons, and never-made bushel, of the English act of 1496, requires of the county commissioners to provide the *avoirdupois* weights and the *Winchester* measures of the English Exchequer.

In the year 1734, a new and amendatory act " for more effectual " obliging persons to buy and sell by weights and measures accord- " ing to the English standard," repeated all the principal provisions of the act of 1661, omitting, however, all reference to the English act of parliament of 1496. And since the Revolution, by an act of the legislature of Virginia, of 26th December, 1792, this act of 1734 is continued, to remain in full force until the Congress of the United States shall have otherwise provided.

Among the numerous wise and honorable examples, which the commonwealth of Virginia has given to her sister states of this Union, has been that of an undertaking to compile and publish a complete collection of her *statutes at large;* that is, of all the acts of her legislative assemblies, from the first settlement of the colony to the present time. This work is, at this time, in the process of publication; and, besides exhibiting the series of all the direct proceedings for the regulation of weights and measures, contains a mass of information, shedding light on every portion of our national history. The connection of weights and measures with the successive progress of this legislation, is more intimate and remarkable from the fact, that the original staple commodity of the colony, *tobacco*, was, for more than a century, not only merchandise, but *money*. It was the circulating medium of exchange; and to a great degree so continued, until supplanted by the modern and less valuable article of bank paper. To trace the varieties of value, affixed to this article of tobacco, in its character of a circulating medium, as rated by legislative enactments, in comparative estimation with other articles of traffic, with the sterling currency of the mother country, with foreign coins of gold, silver, and copper, with the assessment of taxes, the levies of imposts, the wages of labor, and the compensations for public service, would be an inquiry into facts of high and interesting curiosity, but too far transcending the immediate objects of Congress, to be properly comprised in this report. It must suffice to say, that the inspection laws relating to this article have been so numerous and so variant, that the collection of them would alone fill several volumes. The latest of these laws, and that which is now in force, is of 6th March, 1819, and provides, that the tobacco hogshead shall not be more than 54 inches long of the stave, nor more than 34 inches at the head within the crow, making reasonable allowance for prizing, not exceeding two inches above the gage in the prizing head; and that it shall contain 1,250 pounds nett of tobacco, with certain allowances for shrinkage.

The assize of casks for other articles, as in most of the other states in the Union, is regulated by different laws, adapted to the different articles. The barrel for tar, pitch, and turpentine, by an act of 26th December, 1792, must contain $31\frac{1}{2}$ wine gallons, the precise nominal dimensions of the old English wine barrel or half hogshead, as prescribed by acts of parliament time out of mind. But, by the same act of 1792, barrels for beef and pork are to contain 204 pounds nett of meat, with an allowance of $2\frac{1}{2}$ per cent. for shrinkage; and are to be of capacity from 29 to 31 gallons. By another act of 28th December, 1795, barrels for fish are to be of not less than 30, nor more than 32 gallons. By an act of 8th January, 1814, the barrel of *salt* is to contain five bushels; agreeing thereby with the primitive Virginian corn barrel of 1631. But an act of 18th February, 1819, now requires that the barrels for bread, flour, or Indian meal, should be made of staves 27 inches long, and be of $17\frac{1}{2}$ inches diameter at the head, and contain 196 pounds of flour or meal.

The size of staves and heading is regulated by an act of 21st February, 1818, as follows:

Staves—long butt, from 5 feet 6 inches to 5 feet 9 inches long,
from 5 to 6 inches broad,
from 2 to 2½ inches thick,

Short butt and pipe } from 4 feet 6 to 4 feet 9 inches long,
from 3 to 4 inches broad,
from ⅝ of an inch to 1¼ thick,

Hogshead—from 3 feet 6 to 3 feet 9 inches long,
from 3 to 4 inches wide,
from ¾ to 1¼ inch thick,

Barrel—from 2 feet 8 to 2 feet 10 inches long,
not less than 3 inches wide,
not less than ¾ of an inch thick in any place,

Heading—of 28, 30, 32, in due proportion, and not more than 34 inches long,
from 5 to 7 inches broad, dressed and clean of sap, and from ¾ to 1¼ inch thick.

NORTH CAROLINA.

The only law of this state relating to weights and measures, a knowledge of which has been obtained, was enacted prior to the American revolution, during the administration of Governor Gabriel Johnston, and is yet in force. It prohibits the use, in trade, by all the inhabitants or traders within the province, of any weights and measures other than are made and used *according to the standard in the English Exchequer,* and the statutes of England in that case provided. It charges the justices of the county courts to provide, at the charge of each county, sealed weights of half hundred, quarter of hundred, seven pounds, four pounds, two pounds, one pound, and half pound; measures of ell and yard, of brass or copper, measures of half bushel, peck, and gallon, of dry measure, and a gallon, pottle, quart, and pint, of wine measure. It prescribes the appointment of standard-keepers in each county, to whom all weights and measures of the inhabitants are to be brought to be sealed, and who are to be sworn to the faithful discharge of their duties: and it subjects to suitable penalties the various offences of falsifying weights and measures, or of trading with such as have not been duly tried by the standard and sealed. It also repeals all former laws of the province upon the subject.

SOUTH CAROLINA.

By an act of 12th April, 1768, the public treasurer was required to procure, of brass or other proper metal, one weight of 50 pounds, one of 25 pounds, one of 14 pounds, two of 6 pounds, two of 4 pounds, two of 2 pounds, and two of 1 pound, avoirdupois weight, according *to the standard of London;* and one bushel, one half bushel, one peck, and one half peck measures, according to *the standard of London*. The weights were to be stamped or marked in figures denominating their weight, and to be kept by the public treasurer: and by these weights and measures, declared to be the standards, all others in the province were to be regulated. By another act, of 17th March, 1785, subsequent to the Revolution, the justices of the county courts were authorized to regulate weights and measures within their respective jurisdictions, and to enforce the observance of their regulations by adequate penalties.

GEORGIA.

An act of the state legislature of 10th December, 1803, declares the standard of weights and measures established by the corporations of the cities of Savannah and Augusta to be the fixed standard of weights and measures within the state; and that all persons buying and selling shall use that standard until the Congress of the United States shall have made provision on that subject. It directs the justices of the inferior courts, in the respective counties, to obtain standards conformable to those of the corporation of one of those cities; and prescribes regulations for keeping the standards, and for trying, marking, and sealing, by them, the weights and measures of individuals, with penalties for using, in traffic, any others not corresponding with them.

An ordinance of the city council of Augusta directs that all weights for weighing any articles of produce, or merchandise, shall be of the *avoirdupois* standard weights; and all measures for liquor, whether of wine or ardent spirits, of the *wine* measure standard; and all measures for grain, salt, or other articles usually sold by the bushel, of the dry, or *Winchester* measure standard. And it prohibits the use of any other than brass or iron weights, thus regulated, or weights of any other description than those of 50, 25, 14, 7, 4, 2, 1, $\frac{1}{2}$, $\frac{1}{4}$, pound, 2 ounces, 1 ounce, and downwards.

KENTUCKY.

An act of the legislature, of 11th December, 1798, reciting in its preamble that Congress are empowered by the federal constitution to fix the standard of weights and measures, and that they had not passed any law for that purpose, recognizes, as thereby remaining in force within that commonwealth, the act of the General Assembly of Virginia, of the year 1734.

It therefore authorizes and directs the governor to procure one set of the weights and measures specified by the Virginian act of 1734, with measures of the length of one foot and one yard; and declares that the bushel dry measure shall contain $2150\frac{2}{3}$ solid inches, and the gallon of wine measure 231 inches. It provides that these standards shall be kept by the secretary of state of the commonwealth; that the governor shall cause to be made and transmitted to each county, scales and standards conformable to those of the state, which are to be kept by persons to be appointed by the county courts, and with which all the weights and measures, used in trade by individuals, are to be made to correspond.

TENNESSEE.

From a communication received from the governor of the state of Tennessee, it appears that there is in that state no standard of weights and measures fixed by the legislature.

OHIO.

The only act of the legislature of the state of Ohio, on this subject, is of 22d January, 1811. It directs the county commissioners of each county in the state to cause to be made one half bushel measure, to contain $1075\frac{2}{10}$ solid inches, which is to be kept in the county seat, and to be called the standard.

LOUISIANA.

Before the accession of Louisiana to the union of these states, the weights and measures used in the province were those of France, of the old standard of Paris. An account of these, and of the present state of the weights and measures in the state of Louisiana, is submitted in the appendix to this report.

By an act of the legislature of 21st December, 1814, the governor of the state was required to procure, at the expense of the state, weights and measures corresponding with those used by the revenue officers of the United States, together with scales and a seal, to be deposited in the custody of the secretary of the state, to serve as the general standard for the state.

Provision was also made by the same act for the appointment of an inspector at New Orleans, and for furnishing standards to the several parishes throughout the state.

By the last section of this act, a special dry measure is ordained, by the name of a barrel, to contain three and a quarter bushels, according to the American standard, and to be divided in half and quarter barrel. The capacity of this measure, containing, according to the law, 6988.86 cubic inches, is referrible to none of the usual dry measures of the ancient Paris standard; but corresponds with tolerable exactness with the ancient Bordeaux half-hogshead, and with the assize of barrels prescribed by almost all the states of the Union, for packing beef, pork, and flour, for exportation.

INDIANA.

An act of the territorial legislature, of 17th September, 1807, authorized the courts of common pleas of the respective counties in the territory, whenever they might think it necessary, to procure a set of measures and weights for the use of the county; namely, one measure of one foot, or twelve inches English measure, so called; one measure of three feet, or thirty-six inches English measure; one half bushel for dry measure, to contain $1075\frac{1}{2}$ solid inches; one gallon measure, to contain 231 solid inches; the measures to be of wood, or any metal, as the court may think proper; also, one set of avoirdupois weights, to be sealed with the name or initial letters of the county. These weights and measures were to be kept by the clerks of the county courts, for the purpose of trying and sealing those used in their counties. After due notice given by the courts that these standards had been procured, all persons were prohibited from buying or selling by weights or measures not corresponding with them: and the clerk was to try and seal all weights or measures brought to him therefor corresponding with the standard. This act was to continue in force till Congress should otherwise provide.

The provisions of this act are, in substance, and nearly to the letter, repeated in an act of the state legislature, of 21st January, 1818.

There is also an act of 24th December, 1816, regulating the inspection of tobacco; and one of 2d January, 1819, regulating the inspection of flour, beef, and pork. The assize of hogsheads and of casks, prescribed in them, is the same as that of the Virginia laws.

MISSISSIPPI.

An act of the territorial legislature, of 4th February, 1807, directed the treasurer to procure a set of the large avoirdupois weights, according to the standard of the United States, if one were established, but if there were none such, according to the standard of London, with proper scales for weights; together with measures of foot and yard, dry measures of capacity, and liquid *wine* measures. He was also required to furnish each county in the territory with a set of weights, scales, and measures, conformable to the above standards, to be kept by a person appointed by the county courts, under oath, and accessible to all persons desirous of having their weights and measures tried and sealed. Penalties were also annexed to the use of weights and measures not corresponding with these standards.

A subsequent act, of 23d December, 1815, further required of the treasurer to procure six sets of the weights and measures as above described, and to distribute them at suitable places in the several counties of the territory; and additional penalties were prescribed for the use of weights and measures not corresponding with the standard.

An act of the legislature of the state of Mississippi, of 6th February, 1818, " to provide for inspections, and for other purposes," contains many other regulations for the keeping of the standard weights and measures, and for securing conformity to them. It makes no alteration of the standard, but confirms, " until Congress shall fix a

" standard for the United States," that which had already been established. It also requires that barrels of flour should contain 196 pounds nett; and barrels of pork and beef 200 pounds nett of meat.

ILLINOIS.

The territorial act of 17th September, 1807, passed while the state of Illinois formed a part of the Indiana territory.

But by an act of the legislature of this state "regulating weights and measures," of 22d March, 1819, the county commissioners of each county in the state were required to procure, at the expense of the county, one foot and one yard English measure; a gallon liquid or wine measure, to contain 231 cubic inches; corresponding quart, pint, and gill measures, of some proper and durable metal; a half bushel dry measure, to contain eighteen quarts, one pint, and one gill, wine measure, or 1075.2 cubic inches, and a gallon dry measure, to contain one-fourth part of the half bushel, these two measures to be of copper, or brass; also, a set of weights, of one pound, one half pound, one eighth pound, and one sixteenth pound, made of brass or iron, the integer of which to be denominated one pound avoirdupois, and to equal in weight 7,020 grains troy, or gold weight. These weights and measures are to be kept by the clerk of the county commissioners, for trying and sealing the measures and weights in common use.

All persons are authorized to have their weights and measures tried by the standards, and sealed; and are forbidden, upon suitable penalties, to buy or sell by others not corresponding with them.

The most remarkable peculiarity of this act is, its departure from the English standard weights by fixing the avoirdupois pound at 7,020 instead of 7,000 grains troy.

ALABAMA.

This state having formed a part of the Mississippi territory, previously to the admission of the state of Mississippi into the Union in 1817, the acts of that territory of 4th February, 1807, and 23d December, 1815, embraced this section of territory. No act of the state legislature of Alabama, on this subject, is known to have been passed.

MISSOURI.

The territorial legislature, by an act of 28th July, 1813, directed the several courts of common pleas within the territory to provide, for and at the expense of the respective counties, one foot and one yard English measures; one half bushel, to contain $1075\frac{1}{5}$ solid inches, for dry measure; one gallon, to contain 231 solid inches, and smaller liquid measures in proportion; to be of wood, or any metal the court should think proper; also, one set of avoirdupois weights, and one seal, with the initial of the county inscribed thereon: all to be kept by the clerks of the courts of common pleas, or circuit courts, for the purposes of trying and sealing the measures and weights used in their counties.

The use, or keeping to buy or sell, of weights or measures not corresponding with these standards, after due notice, was prohibited under penalties by the same acts; but with a proviso, that all contracts or obligations, made previous to the taking effect of the act, should be settled, paid, and executed, agreeably to the weights and measures in common use when the contracts or obligations were made or entered into.

DISTRICT OF COLUMBIA.

By the act of Congress of 27th February, 1801, concerning the District of Columbia, the laws of the state of Virginia, as they then existed, were continued in force in the part of the District which had been ceded by that state, and the laws of Maryland in the part of the District ceded by Maryland.

The act to incorporate the inhabitants of the city of Washington, of 3d May, 1802, authorizes the corporation to provide for the safe keeping of the standard of weights and measures fixed by Congress, and for the regulation of all weights and measures used in the city.

The supplementary act, of 24th February, 1804, gives the city council power to establish and regulate the inspection of flour, tobacco, and salted provisions; and the gauging of casks and liquors.

And by the act of 4th May, 1812, further to amend the charter of the city of Washington, further power is given to the corporation to regulate the measurement of, and the weight by which, all articles brought into the city for sale shall be disposed of.

The weights and measures of the city have, accordingly, been regulated by various acts of the corporation, conformably to the standard used in the state of Maryland. The inspection laws, the assize of tobacco hogsheads and flour casks, the dimensions of bricks and of cord wood, are all formed upon the same model. The weight of bread is adapted once a month to the price of flour: but by a special ordinance, all coal for sale within the city is sold by a measure containing five struck-standard half bushels, stamped and marked by the sealer of weights and measures, and the stricken measure of which is considered as two bushels.

As preliminary remarks, in reference to that part of the resolutions of both Houses, which requires the opinion of the Secretary of State with regard to the measures which it may be proper for Congress to adopt in relation to weights and measures, it may be proper to state the extent of what can be done by Congress. Their authority to act is comprised in one line of the constitution, being the fifth paragraph of the eighth section and first article; in the following words: "*to fix the standard of weights and measures.*"

It may admit of a doubt whether under this grant of power is included an authority so totally to subvert the whole system of weights and measures as it existed at the time of the adoption of the constitutution, as would be necessary for the introduction of a system similar to that of the French nation. To *fix* the standard, appears to be an operation entirely distinct from changing the denominations and proportions already existing, and established by the laws, or immemorial usage. And this doubt acquires a further claim to consideration, if it be true, as the experience of other nations seems to warrant us in the conclusion, that there is no object of regulation by human power, in which the prescriptions of a government are so difficult to be carried into execution. Throughout Europe, in the most absolute as well as in the freest governments, every historical research presents a fruitless struggle on the part of authority to introduce order and uniformity: and an unconquerable adherence of custom to the diversities of usage among the people. There is perhaps less of this diversity in the United States, than in any country in Europe. At the adoption of the constitution all the weights and measures in common use throughout the United States were derived, either by the statutes of the states, or by an invariable usage, which had supplied the place of law, from the standards in the English exchequer. Hence, the English foot, divided into twelve inches, was the unit of all measures of matter in length, breadth, or thickness. Its various multiples of the yard, ell, perch, pole, furlong, acre, and mile, were all recognized by the laws, and in the familiar use of the people. The avoirdupois and troy weights with the difference of modification of the latter as used for weighing the precious metals or apothecary's drugs in retail, the wine gallon of 231, and the beer gallon of 282 solid inches, were equally well known, and in general use, and the Winchester bushel, of 2150.42 solid inches, formed the general standard of all the dry measures of capacity.

In many of the states the standards established by statute had been procured from the court of exchequer; and the only variety discernible in the legislation of the states on this subject, arises from a difference existing in the several standards of the same measures at the exchequer, and at Guildhall in London.

In the exercise of the authority of Congress, with a view to the general principle of uniformity, there are four different courses of proceeding which appear to be practicable.

1. To adopt, in all its essential parts, the new French system of weights and measures, founded upon the uniformity of identity.

2. To restore and perfect the old English system of weights, measures, moneys, and silver coins, founded upon the uniformity of proportion.

3. To devise and establish a system, in which the uniformities of identity and of proportion shall be combined together, **by adaptations of parts of each system to the principles of the other.**

4. To adhere, without any innovation whatever, to our existing weights and measures, merely fixing the standard.

1. In the review which has been taken, and the comparison which has been submitted to Congress, between the old English, and the new French, or as they may with more propriety be called, the ancient and the modern systems of metrology, it has been the endeavor of this report to show, that, while each of these systems embraces principles of the highest importance, neither of them includes all the elements resulting from the nature of the relations between man and things as created beings, and between man and man in society, mingling in the purposes to which weights and measures are applicable. The opinion has been expressed, that the uniformity of proportion in the ancient system, uniting weight and measure by the relative gravity, extension, and numbers, incident to dry and liquid substances, possessed advantages, of which the uniformity of identity in the modern system was entirely deprived; that the property of the ancient system, by which the money weight and the silver coin were the same, the most useful of all uniformities of which weights, measures, money, and coins are susceptible, was very imperfectly adapted to the modern system of France; that the French system, admirable as it is, looked, in its composition, to weights and measures, more as exclusively matters of account, than as tests of quantity; that, in its eagerness for extreme accuracy in the relations between things, it lost sight a little of the relations of weights and measures with the physical organization, the wants, comforts, and occupations of man; that, in its exclusive partialities for decimal arithmetic, it forgot the inflexible independence and the innumerable varieties of the forms of nature, and that she would not submit to be trammelled for the convenience of the counting house. The experience of the French nation under the new system has already proved, that neither the immutable standard from the circumference of the globe, nor the isochronous vibration of the pendulum, nor the gravity of distilled water at its maximum of density, nor the decimation of weights, measures, moneys, and coins, nor the unity of weight and measure of capacity, nor yet all these together, are the only ingredients of practical uniformity for a system of weights and measures. It has proved, that gravity and extension will not walk together with the same staff; that neither the square, nor the cube, nor the circle, nor the sphere, nor the revolutions of the earth, nor the harmonies of the heavens, will, to gratify the pleasure, or to indulge the indolence of man, be restricted to computation by decimal numbers alone.

The substitution of an entire new system of weights and measures, instead of one long established and in general use, is one of the most arduous exercises of legislative authority. There is indeed no difficulty in enacting and promulgating the law; but the difficulties of carrying it into execution are always great, and have often proved insuperable. Weights and measures may be ranked among the necessaries of life, to every individual of human society. They enter into the economical arrangements and daily concerns of every family.

They are necessary to every occupation of human industry; to the distribution and security of every species of property; to every transaction of trade and commerce; to the labors of the husbandman; to the ingenuity of the artificer; to the studies of the philosopher; to the researches of the antiquarian; to the navigation of the mariner, and the marches of the soldier; to all the exchanges of peace, and all the operations of war. The knowledge of them, as in established use, is among the first elements of education, and is often learnt by those who learn nothing else, not even to read and write. This knowledge is rivetted in the memory by the habitual application of it to the employments of men throughout life. Every individual, or at least every family, has the weights and measures used in the vicinity, and recognized by the custom of the place. To change all this at once, is to affect the well-being of every man, woman, and child, in the community. It enters every house, it cripples every hand. No legislator can attempt it with any prospect of success, or any regard to justice, but upon two indispensable conditions: one, that he shall furnish every individual citizen easy access to the new standards which take the place of the old ones; and the other, that he shall enable him to *know* the exact proportion between the old and the new. A multiplication of standard copies to a great extent is indispensable; and the distribution of them throughout the country, so that they may be within the means of acquisition to every citizen, is among the duties of the government undertaking so great a change. Tables of equalization must be circulated in such a manner as to find their way into every house; and a revolution must be effected in the use of books for elementary education, and in all the schools where the first principles of arithmetic may be taught. All this has been done in France; and all this might be done perhaps with more ease in the United States. But, were the authority of Congress unquestionable to set aside the whole existing system of metrology, and introduce a new one, it is believed that the French system has not yet attained that perfection which would justify so extraordinary an effort of legislative power at this time.

The doubts entertained whether an authority, so extensive as this operation would require, has been delegated to Congress, are strengthened by the consideration of the character of the executive power, corresponding with the legislative authority. The means of execution for exacting and obtaining the conformity of individuals to the ordinances of the law, in the case of weights and measures, belong to that class of powers which, in our complicated political organization, are reserved to the separate states. The jurisdictions to which resort must be had for transgressions of this description of laws, are those of municipal police. In England they were originally of the resort of views of frankpledge in every separate manor, and have since been transferred to the clerks of the market and to the justices of the peace. The sealers of weights and measures, officers who have the custody of the standards, and the authority to compare with them, from time to time, the weights and measures used by individuals, and to prosecute for all offences by variations from the standards, and the courts before whom all such offences are triable, are institutions not only existing in almost every state in the Union, but essentially belonging to that portion of public authority suited to the state administration rather than to that of the Union. It is a general principle of our constitutions, that, with every delegation of legislative authority, a co-extensive power of execution has been granted. Affairs of municipal and domestic concern have, for obvious reasons, been reserved to the state authorities ; and of this character are most of the regulations and penal sanctions for securing conformity to the standards of weights and measures. In *fixing the standard*, it is believed that Congress must rely almost entirely, if not altogether, upon state executive authorities, for carrying their law into execution. And, although this reliance may be safely indulged in relation to a law which should merely fix the uniformity of existing standards, its efficacy would be very questionable in the case of a law of great and universal innovation upon the habits and usages of the people. Of such a law the transgressions could not fail to be numerous : any doubt of the authority of the legislator would stimulate to systematic resistance against it : and the power of enforcing its execution being in other hands, naturally disposed to sympathise with the offender, the whole system would fall into ruin, and afford a new demonstration of the impotence of human legislation against the laws of nature, in the habits of man.

2. The restoration of the old English, which was also the Greek and Roman, system of weights, measures, and silver coins, founded upon the uniformity of proportion, would require an exercise of authority no less transcendent than the introduction of the French system. Its advantages were, the identity of the money weight and silver coin, the wine gallon at once a multiple of the money weight, and an aliquot part of the cubic foot ; and its proportions of the money and commercial pounds, and of the wine and corn gallons, to the relative specific gravity of wine and wheat. But, as all these combinations were founded upon the assumption that the relative gravity of wheat to wine was as 4 to 5, and that the gravity of wine and of spring water was the same; and as it allowed of the making of the wine gallon by the two processes, by the weight of wheat multiplied, and by the weight of the cubic foot of water divided, the result of the two processes was not exactly the same. The Irish gallon, of 217.6 inches, was made by one process ; and the Rumford gallon, of 266.25, was its corresponding corn and ale measure. A wine gallon of 219.5 cubic inches was made by assuming 252 gallons as the measure of the ton, or 32 cubic feet ; and its corresponding corn measure was the Winchester bushel, with an ale gallon of 268. The Winchester bushel is the only existing relic of the old English system, which has outlived all the changes of the laws, and all the revolutions of ages. Should that be retained, and its contents fixed at 2148.5, to restore and perfect the whole system by an exact combination of the two modes of forming the water gallon, without regard to the weight of

wine, would require a liquid gallon of 219.5 inches, a dry gallon of 268.5, a money pound of 5714.28, and a commercial pound of 6944.44 grains troy. This money pound should then be made the weight of the unit of silver coins, of a settled standard purity, and might be decimally divided, like our present silver coins, and decimally or duodecimally divided as a weight. Or, the ton might be declared to contain 256 gallons, of 216 cubic inches; in which case the money pound would be of 5625, and the commercial pound of 6836 grains troy; the corn and ale gallon of 262.5, and the bushel of 2100 cubic inches. If the old easterling 12 and 15 ounce pounds should be restored, and the gallon, according to its primitive composition, be made to contain ten 12 ounce pounds of wine, it would then be, considering the gravity of wine as of 250 grains troy to a cubic inch, of the same capacity of 216 cubic inches. It would also contain eight 15 ounce pounds, of 6750 grains troy; but the proportion between the two pounds would not be exactly that between the gravity of wheat and wine. The wine gallon, filled with eight 12 ounce pounds of wheat, would contain, in wine, eight pounds, not of 6750, but of 6608 grains; and, if divided into fifteen ounces, the ounce would not be the easterling, but the avoirdupois ounce.

3. The proportions between the existing troy and avoirdupois weights, and between the wine gallon of 231, and the beer gallon of 282 cubic inches, are more exactly those between the specific gravity of wheat and of spring water, than were the easterling pounds of 12 and 15 ounces, or those of the primitive gallon of 216 inches with the ale gallon deduced from the Winchester bushel. They are exact, to the utmost degree of precision; but these proportions are without use. Neither does the wine gallon contain an exact number of pounds of wine, nor is the beer gallon an aliquot part of the bushel. These were proportions, in their origin, of great usefulness, but imperfectly settled. The whimsical operation of time and human laws upon them has been to make the proportions perfect, but to render them useless. There are, nevertheless, very useful proportions in our existing weights and measures, one of which is between the ton measure of water and the pound avoirdupois. As 1,000 ounces avoirdupois weigh exactly one cubic foot of water, it follows that the ton of 2,000 pounds weight is the ton of 32 cubic feet measure. The other is between the pound avoirdupois and the pound troy; the former consisting of precisely 7,000 grains troy. The pound avoirdupois is therefore the connecting link between weight and linear measure. It is at once a test and standard of the cubic foot, of the ton measure, and of the troy weight; while the foot, the ton, and the troy weight, are each, by this connecting link, tests and standards of each other, and of the avoirdupois pound. But the thirty-two cubic feet, which are at once the ton weight of two thousand pounds, and the ton measure of water, are not sufficient, as measure, to contain the same weight of wheat. The bushel is the measure containing the same weight of wheat which the cubic foot contains of water. Thirty-two bushels, therefore, contain the ton weight, of two thousand pounds avoir-

dupois: but they would make a ton measure within a small fraction of 39 cubic feet.

The avoirdupois pound of 16 ounces, and of 7,000 grains troy, is used, however, only for quantities of less than a quarter of a hundred pounds. It then receives an accession of 12 per cent. on its quantity: the quarter of a hundred contains 28 pounds, the hundred 112, and the ton of 2,000 actually contains 2,240. If the hundred and twelve pounds should be considered as a nett hundred, each pound would be of 7,840 grains troy weight, and would bring it within one-quarter of an ounce troy to the weight of the French half kilogramme or usual pound. If the wine gallon were, as under the statute of 1496 it should have been, and as the Guildhall gallon before the statute of 5 Anne actually was, of 224 inches, it would have had two further useful coincidences: it would have contained just eight pounds avoirdupois of wine; eight pounds troy weight of wheat; and a number of cubic inches in decimal subdivision to the number of pounds avoirdupois in the ton of 2,240, or twenty hundred of 112 pounds.

There are two changes, therefore, in our existing weights and measures, which would restore and perfect the system of ancient metrology; one, to make the troy weight the unit of our silver coins, in which case it might be decimally divided as coin, retaining its divisions into ounces, pennyweights, and grains, as a weight; and the other, to restore the wine gallon of 224 inches, with its corresponding ale gallon of 272, and bushel of 2,176 inches.

But it has been already remarked, that in the ancient system, founded on the uniformity of proportion between the relative extension and gravity of wheat and wine, there were, in the double sets of weights and measures of capacity, two advantages; one, of a general nature, resulting from it as proportional, without reference to the articles selected for settling the proportions; and the other special, arising from the selection of wheat and wine as the articles. The first belongs to every proportional system, of which the proportion between the standards is accurately ascertained, and consists in this, that each weight and each measure is a test and standard for all the others. The second depends on the selection of the articles, and is limited to the conveniences and facilities of trade, commerce, and navigation, as incidental to them. Reasons have been suggested, why the two articles of wheat and wine should have been selected in the primitive system, as being, from the nature and physical constitution of man, the first, and, for many ages, the greatest and most important articles of traffic. The necessity for establishing a proportion between the relative weight and measure of those articles, was also dictated by the practice of transporting them both by sea in ships.

The space in cubic feet which would be filled by a determinate weight of each of them was an object of essential importance to be known, not as a philosophical theory, but for every mechanical operation of the commerce. The size of the cask must be adapted to the capacity and the burthen of the ship: and when the ton weight of wine had been adapted to the ton measure of water, it became of the utmost use to make the measure of corn so correspond with the cask of wine, as to contain the same determinate quantities by weight. But in modern times, and especially to these United States, neither wheat nor wine is an article of primary importance in domestic trade, or in foreign commerce. Whatever may be the capacities of our country for producing wine, they have hitherto scarcely been discovered. Tea and coffee have taken the place of wine as comforts, or next to necessaries of life; and have degraded that article into the class of luxuries. We import little, and export none of it. We receive it in the casks of the several countries from which it comes: and although the laws of some of our states, as well as those of England, still exhibit the absurdity of requiring that the hogshead should contain 63 wine gallons of 231 cubic inches, because it once contained 63 gallons of 219½ inches, yet no one complains that the real hogshead is just what it was 600 years ago, without either swelling to the dimensions of queen Anne's cubic inches, or contracting the gravity of its contents to the troy weight of Henry the Seventh. We raise vast quantities of wheat, but export it almost exclusively in its manufactured state of flour. The weight of wine is, between the buyer and seller, never a subject of inquiry. We have universally the Winchester bushel, defined by the 13 William III, of 2,150.42 cubic inches, with the single exception of the state of Connecticut, whose standard bushel is *very near* 2,198 inches. And the laws of many of the states require, that the bushel should contain 60 pounds avoirdupois of wheat. Should a standard bushel now be made in the manner described in the statute of 1266, it would be a measure of 2,148.5 cubic inches, and would contain 60½ avoirdupois pounds of wheat. The relative proportion between the extension and specific gravity of wheat and wine is to us, therefore, of no importance or use in our system of weights and measures. When the wine gallon contained a determinate weight of the liquor, and was at the same time a 63d part of eight cubic feet, there were motives of convenience and utility in using another measure for ale and beer, which, being brewed from grains, had natural proportions to the measures used for them. It was natural, therefore, to employ the eighth part of the measure of the bushel as the beer gallon, though at the same time a vessel of smaller size was used for the measurement of wine. But since the weight of wine, and the proportions of its measuring vessel to the cubic foot, have ceased to be of any account, there is no purpose of utility answered by the employment of two different measures for different fluids; while there is great tendency to error and fraud in the use of two such measures, of the same materials and bearing the same name.

4. Our system of weights and measures is, therefore, susceptible of great improvements, by restoring some of the principles which belonged to the system from which it was originally derived. It is perhaps still more improvable, by the adoption of some of the principles contained in the new French metrology. There is no doubt that the decimal divisions might be introduced to great advantage

both into linear measure by the adoption of the metre, and into weights, by identifying the money weight with the silver coin. It is believed that a system, embracing the essential advantages of all the three, might, without much difficulty, be combined; and that it would be better adapted than either of them, to the use of all human kind, and thus secure, in its utmost possible extent, the uniformity with reference to persons.

Weights and measures, and the final establishment of a system for them, with a view to the utmost practicable extent of uniformity, are at this moment under the deliberative consideration of four populous and commercial nations—Great Britain, France, Spain, and the United States. The interest is common to them all: the object of *uniformity* is the same to all. Could they agree upon one result, the advantages of that agreement would be great to each of them separately, and still greater in all their intercourse with one another. But this agreement can be obtained only by consultation and concert. It is, therefore, respectfully proposed, as the foundation of proceedings necessary for securing ultimately to the United States a system of weights and measures which shall be common to all civilized nations, that the President of the United States be requested to communicate, through the ministers of the United States, in France, Spain, and Great Britain, with the governments of those nations, upon the subject of weights and measures, with reference to the principle of uniformity as applicable to them. It is not contemplated by this proposal, that the communication should lead to any conventional stipulations or treaties; but it is hoped that the comparison of ideas, and the mutual reciprocation of observation and reflection, may terminate in concurrent acts, by which, if even universal uniformity should be found impracticable, that which would be obtained by each nation would at least approximate nearer to perfection.

In the mean time, should Congress deem it expedient to take immediate steps for accomplishing a more perfect uniformity of weights and measures within the United States, it is proposed that they should assume as their principle, that no innovation upon the existing weights and measures should be attempted.

To fix the standard of weights and measures of the United States as they now exist, it appears that the act of Congress should embrace the following objects:

1. To *declare* what are the weights and measures to which the laws of the United States refer as the legal weights and measures of the Union.

2. To procure positive standards of brass, copper, or such other materials as may be deemed adviseable, of the yard, bushel, wine and beer gallons, troy and avoirdupois weights; to be deposited in such public office at the seat of government as may be thought most suitable.

3. To furnish the executive authorities of every state and territory with exact duplicates of the national standards deposited at the seat of government.

4. To require, under suitable penal sanctions, that the weights and measures used at all the custom-houses, and land surveys, and post offices, and, generally, by all officers under the authority of the United States, in the execution of their laws, should be conformable to the national standards.

5. To declare it penal to make or to use, with intent to defraud, any other weights and measures than such as shall be conformable to the standards.

1. The existing weights and measures of all the states of this Union are derived from the exchequer, or from the laws of Great Britain. The one common standard, from which they are all deduced, is the English foot, divided into twelve inches, and three of which constitute the yard. The positive standard yard is a brass rod of the year 1601, in the British exchequer. The unit of measure is the foot of twelve equal inches. The inch, by the English laws, is divided into three equal parts, called barley-corns; but this division is not used in practice. The practical divisions of the inch, are, at option, binary, or decimal; that is, of halves, quarters, and eighths; or of tenths, hundredths, and thousandths. Thirty-two cubic feet of spring water, at the temperature of 56 degrees of the thermometer of Fahrenheit, constitute the ton weight of two thousand pounds avoirdupois. The pound avoirdupois consists of sixteen ounces; the ounce, of sixteen drams. The pound avoirdupois is equal in weight to seven thousand grains troy, or to fourteen ounces, eleven pennyweights, sixteen grains troy. The troy pound consists of twelve ounces; each ounce of twenty pennyweights, each pennyweight of twenty-four grains. It is otherwise divided for the use of apothecaries; but the grain and the pound are the same. The troy pound is equal in weight to 13 ounces and $2\frac{2}{3}$ drams avoirdupois.

The bushel is a cylindrical vessel $18\frac{1}{2}$ inches in diameter, and eight inches deep; or any vessel of 2,150.42 cubic inches. It is divided into four pecks, each peck into four pottles, each pottle into two quarts, each quart into two pints.

The ale and beer gallon is a vessel of 282 cubic inches. It is divided into four quarts, each quart into two pints, each pint into four gills.

The wine gallon is a vessel of 231 cubic inches; divided, like the beer gallon, into wine quarts, pints, and gills.

Any cubic vessel of 12.9 inches in length, breadth, and thickness, is of equal contents with the Winchester bushel. Any cubic vessel of 6.55767, is of equal contents with the ale gallon. Any cubic vessel of 6.13579 is a wine gallon.

For the purposes of the law, it will be sufficient to declare, that the English foot, being one-third part of the standard yard of 1601 in the exchequer of Great Britain, is the standard unit of the measures and weights of the United States; that an inch is a twelfth part of this foot; that thirty-two cubic feet of spring water, at the temperature of 56 degrees of Fahrenheit's thermometer, constitute the ton weight, of 2,000 pounds avoirdupois; that the gross hundred of avoirdupois weight consists of 112 pounds, the half hundred of 56, and the quarter hundred of 28, the eighth of a hundred of 14, and the sixteenth of a hundred of 7 pounds; that the troy pound consists of 5,760 grains, 7,000 of which grains are of equal weight with the avoirdupois pound; that the bushel is a vessel of capacity of 2,150.42 cubic inches, the wine gallon a measure of 231, and the ale gallon a measure of 282 cubic inches.

The various modes of division of these measures and weights, the ell measure, and the application of the foot to itinerary, superficial, and solid measure, producing the perch, rood, furlong, mile, acre, and cord of wood, may be left to the established usage, or specifically declared, as may be judged most expedient. The essential parts of the whole system are, the foot measure, spring water, the avoirdupois pound, and the troy grain.

2. For the purpose of uniformity, it would be desirable to obtain a copy, as exact as the most accomplished art could make it, of the standard yard of 1601, in the exchequer of Great Britain, made of the same material, brass, but divided with all practicable accuracy into three feet, and thirty-six inches, and each inch further divided into tenth and hundredth parts. This rod, with the words, " standard yard measure of the United States—three feet—thirty-six inches:" and the date of the year engraved on one of its sides, should be enclosed in a wooden case, and deposited for safe-keeping in one of the offices at the Capitol. From the foot measure of this yard, the standard bushel, and two gallons, should be made. The avoirdupois pound, and the troy weight of 256 ounces, should be made exactly conformable to the standards in the exchequer. The weights of 56, 28, 14, and 7 pounds avoirdupois, should be made exact multiples of the pound weight. But no subdivisions of the bushel or gallons, or of the avoirdupois pound, should be placed among the standards. An enactment, that no subdivisions of the standards, other than in the due proportion to them, should be legal, would avoid the inconvenience and the varieties which multiplied material standards always produce. All the standards should, like the yard, have their names, as standards of the United States, the date of the year, and a designation of quantity engraved upon them. On the bushel, for instance,—" $2150\frac{42}{100}$ cubic inches;" on the wine and ale gallons, respectively, 231 and 282 inches; on the avoirdupois pound " 7,000 " grains troy weight, avoirdupois pound," on the troy weights " 256 " ounces—12 ounces, and 5,760 grains to the pound troy weights." These standards, all enclosed in suitable cases, to preserve them from injury, and, as effectually as possible, from decay, should be deposited in the custody of a sworn and responsible officer, with the standard yard.

3. These national standards being thus made and deposited, exact copies of them should be made of the same materials, substituting for the words " standard of the United States," engraved upon the originals, the words " United States" standard, state " of———:" and these copies should be transmitted to the executives of every state in

the Union. The standard for the territories might leave the name of the state to be engraved when the territory should pass to that condition: and the standards for the District of Columbia might properly be committed to the charge of the clerk of the supreme court of the United States.

4. It should be made the duty of the collectors, surveyors, and naval officers of the customs, the registers of the land offices, and receivers of public moneys, of the postmaster general, and all postmasters, the quartermasters, and commanding officers at military posts of the army, the commanding officer and purser of every vessel of the navy, the commanding officer at the military academy, of all Indian agents, and of the marshals of the several judicial districts of the United States, to ascertain, and to certify in writing, upon oath, to the heads of their respective departments, that the weights and measures used by them, in the discharge of their official duties, are conformable to the standards of the United States. And to secure the future observance of this uniformity, every such officer, civil or military, to be appointed hereafter, should, together with the oath to support the constitution of the United States, have administered to him an oath that he will, in the discharge of his official duties requiring the employment of weights and measures, scales and beams, use such as are conformable to the legal standards of the United States, and not knowingly any others. To the penalties of removal from, and disqualification for office, might be added a right of action for damages, given to any person injured by the wilful neglect or refusal of any such officer to observe the requisitions of the law.

5. The offence of fraudulently or wilfully making or selling any weight, measure, scales, or beam, to be used as conformable to the United States' standards, and not conformable, might be made punishable by fine and imprisonment, upon presentment and conviction before the circuit courts of the United States.

The existing laws of all the states should be declared, so far as they are conformable to the act of Congress *fixing* the standard, to remain unrepealed and in full force. All sealers of weights and measures, and all persons appointed under the authority of the several states for the custody of standards, should be required to ascertain them to be conformable to the standards of the United States. It is scarcely possible that any law of the United States to establish uniformity of weights and measures throughout the Union, should be made effectual, without the cordial aid and co-operation of the state legislative and executive authorities. This is one of the most powerful reasons which have led to the conclusion, that, in fixing the standard, all present innovation should be avoided. The standards of all the states are now, or by their laws should be, the same as those herein proposed, excepting only the Connecticut bushel, the change in which will be inconsiderable. Several of the states have systems well organized, and in full operation for the uniformity of their weights and measures. The standards of many of them are incorrect; some from careless usage and decay; others from having been copies of copies made without much attention to accuracy ; and, others from having transferred to this country all the varieties of the original standards in the exchequer. The object of the act, the substance of which is now proposed to Congress, would be, to make the uniformity already existing by the laws and usages of every part of the Union more effectual and perfect in point of fact. The table* of a return from the several custom houses of the United States will shew the extent of the existing varieties; and while they add new demonstration of the justness of the sentiment universally prevailing, that the authority delegated to Congress by the constitution, of fixing the standard, should be exercised without delay, they also show that the best exercise of that authority will be by making it essentially auxiliary to the efficacy of the existing state laws.

In the consultation which it is proposed that the President of the United States should be requested to authorize and conduct with foreign governments, with a view to future, more extensive, and perfect uniformity, there is one object, which, it is presumed, may be accomplished with little difficulty or expense, and by means of which the standard from nature of the new French system, the metre, may be engrafted upon our system without discomposing any of its existing proportions.

In all the proceedings, whether of learned and philosophical institutions, or of legislative bodies, relating to weights and measures within the last century, an immutable and invariable standard from nature of linear measure has been considered as the great desideratum for the basis of any system of metrology. It is one of the greatest merits of the French system to have furnished such a standard for the benefit of all mankind, in the metre, the ten millionth part of the quarter of the meridian. Of the labors, and researches, and liberal expense, and art, and genius, which have been lavished by France upon this operation, and of the success with which it has been accomplished, the notice which it amply merited has already been taken in this report. Since this great and admirable undertaking has been achieved, a disposition to detract from its merit and usefulness has been occasionally manifested. Some philosophical speculators have started doubts whether the metre is really the forty millionth part of the circumference of the earth; and indeed whether such a measure can, with perfect accuracy, be ascertained by human art. Other standards from nature have been suggested as preferable to the arc of the meridian: individual passions and antisocial prejudices have insinuated themselves into the inquiry: and the question between the metre and the pendulum has almost festered into a test of party controversy, and an engine of national jealousy. In the establishment of the French system, the pendulum, as well as the meridian, has been measured ; but the *standard* was, after long deliberation, after a cool and impartial estimate of the comparative advantages and inconveniences of both, definitively assigned to the arc of the meridian, in departure

* See Appendix.

from an original prepossession in favor of the pendulum. Two reasons are deemed decisive for concurring in the principle of this determination; one, that the earth being the greatest object of actual measurement within the physical powers of man, an aliquot part of its circumference is the only measure, which, applicable to that object, is also equally applicable to every other purpose of weight or mensuration; and the other, that this standard once settled is invariable, while the pendulum, being of different lengths in different latitudes, is essentially defective in one of the most important principles of uniformity, that of *place* or capacity of application to every part of the earth.

It is proposed, therefore, to discard all consideration of the pendulum: as the theory of its vibrations, however interesting in itself, is believed to be, since the definitive determination of the metre, useless, with reference to any system of weights and measures. Nor is it of more importance to know whether the metre really be, within the ten thousandth part of an inch, an exact aliquot part of the circumference of the earth. An error to that, or even to a greater extent, admitted to be possible, leaves for all practical purposes of human life, even including the operations of geography and astronomy, the metre as perfect a standard for weights and measures as any other that ever was devised, and a much more perfect one than the pendulum.

It is therefore submitted to the consideration of Congress, that, in the act for fixing the standard of weights and measures for the United States, together with a definition of the foot, its exact proportion to the standard metre of France should be declared: to effect which purpose with the utmost attainable accuracy, it would be necessary to compare together the identical measure, to be used hereafter as the standard linear measure of the Union, with the standard metre in platina, deposited in the national archives of France. It is not doubted that the French government would readily give their assent to this operation, and would agree that it should be performed in such manner as to settle, definitively, for the future use of both countries, the exact proportion to the ten thousandth part of an inch, between the foot measure of the United States and the metre. From the perfection which the instruments used for comparing together measures of length have attained, accuracy to that extent may be effected. But the necessity of such an operation for the definitive settlement of this proposition is apparent, from the fact, that the comparisons hitherto made in France, and in England, and in the United States, though all made with all possible care, have terminated in results so different, that it would scarcely be safe to assume either of them as the proportion to be declared by a legislative act.

In the attempt to determine distances of space less than the 200th part of an inch, the experiment is met by obstacles, in the temperature and pressure of the atmosphere, and in the different degrees of their influence upon the matter to be measured. Heat and cold, moist and dry, high and low, affect the metals of which measures are composed with various degrees of dilatation and contraction. Brass, the metal of which the English standards are formed, being a compound metal, is variously dilatable: and, although tables have been formed of the degrees in which the simple metals are expanded by heat, according to the scale of the thermometer, yet, as those tables, made by different men, do not agree, no perfect reliance can be had upon them. As yet no experiments of admeasurement, made by different persons, at different times, but of the same standards, have exhibited results, approximating within one two-hundredth part of an inch : of a contrary result, the examples are numerous, and so remarkable that they deserve to be noticed more particularly.

In the year 1797, sir George Shuckburg Evelyn measured, with Troughton's microscopic beam compass and scale, all the standards at the exchequer; the scale made by Sisson for Graham in 1742, the parliamentary standards of 1758 by Bird, the scale used by general Roy for the measurement of the base, and several others. The result of his experiment was published in the transactions of the Royal Society of that year. He found that the standard yard of Elizabeth at the exchequer marked 36.015 inches, Bird's parliamentary standard of 1758, 36.00023, and general Roy's scale 36.00036, on the scale of Troughton.

In the year 1818, captain Henry Kater, one of the commissioners of the prince regent, with the same microscope beam compass of Troughton, measured the same scale of general Roy, and found 39.4 inches on the latter to be equal to 39.40144 on the scale of Troughton. The difference between these two results is $\frac{105}{10000}$, or rather more than a hundredth part of an inch. Captain Kater, to account for it, supposes, that when sir George Shuckburg made the comparison, the two scales were not at the same temperature: but sir George Shuckburg, in his own account of his experiments, expressly mentions his leaving together another of the scales with that of Troughton, by which he measured it 24 hours, that they might acquire the same temperature ; and marks the state of the thermometer (51.7) when he measured the scale of general Roy. Captain Kater states the thermometer, when he measured it, to have been at 70.

A difference equally striking has happened in the experiments made in France and England, to ascertain the relative proportions of the English foot and of the French metre. The result of numerous experiments, made in France under the direction of the National Institute, or Academy of Sciences, has been to announce the metre to be precisely equal to 39.3824 English inches. The result of captain Kater's experiments, after numerous others under the direction of the Royal Society, is the declaration, that the French metre is equal to 39.3708 English inches. The difference is $\frac{116}{10000}$ of an inch, more than one hundredth part, and as near as possible the same as that of the experiments of captain Kater and of sir George Shuckburg Evelyn, upon the scales of Troughton and of general Roy.

A very interesting account of experiments made in this country by Mr. Hassler, to ascertain the length of the metre, is subjoined to this report, from which the mean length of four standard metres was found to be 39.38024797 English inches upon a scale of Troughton's, of equal perfection with that of sir George Shuckburg Evelyn.

Again ; in the year 1814, the committee of the house of commons resolved, and, in 1815, the house itself enacted, that the length of a pendulum vibrating seconds, in the latitude of London, had been ascertained to be 39.13047 inches of Bird's parliamentary standard yard.

In the year 1818, captain Kater reported, as the result of his experiments, that the length of the pendulum vibrating seconds *in vacuo* at the level of the sea, at the temperature of 62° of Fahrenheit, in latitude 51° 31' 8" 4''' north, (London) was 39.13842 inches of the same Bird's parliamentary standard yard.

The difference is $\frac{795}{10000}$, or a one hundred and twenty-sixth part of an inch.

By assuming a mean average from all these experiments, and the yard of Elizabeth at the exchequer (the standard from which all the long measures of the United States are derived,) as the measure of comparison, we might be warranted in taking 39.38 English as the length of the French platina standard metre, and 39.14 as the length of the pendulum vibrating seconds in the latitude of London. And if the attempt at a minuter decimal fraction than that of the 100th part of an inch in the making of metallic measures, should terminate again in disappointment, it is nevertheless true, that, to obtain accuracy even to that extent, the microscopic beam compasses, and the micrometer marking subdivisions to the 25,000th part of an inch, are essential auxiliaries: for in this, as in all the energies, moral or physical, of man, the pursuit of absolute perfection is the only means of arriving at the nearest approximation to it, attainable by human power.

When the proportion shall be thus ascertained, by a concurrent agreement with France, the act might declare that the foot measure of the United States is to the standard platina metre of France in such proportion that 39.3802 inches are equal to the metre, and that 472.5623 millimetres are equal to the foot. The proportion of the troy and avoirdupois pounds to the kilogramme might be ascertained with equal accuracy, and declared in like manner. A platina metre and kilogramme, being exact duplicates of those in the French national archives, should then be deposited and preserved with the national standards of the United States.

It is not proposed that the standard yard measure of the United States should be made of platina ; but that it should be of the same metal as the yard of 1601, at the Exchequer, from which it will be taken. The very extraordinary properties of platina, its unequalled specific gravity, its infusibility, its durability, its powers of resistance against all the ordinary agents of destruction and change, give it advantages and claims to employment as a primary standard for weights and measures, and coins, to which no other substance in nature has equal pretensions. The standard metre and kilogramme of France are of that metal. Should the fortunate period arrive when the improvement in the moral and political condition of man will admit of the introduction of one universal standard for the use of all mankind, it is hoped and believed that the platina metre will be that measure. But, as the principle respectfully recommended in this report is that of excluding all innovation or change, for the present, of our existing weights and measures, it is with a view to uniformity that the preference is given, for the choice of a new standard, to the same metal of which that measure consists which has been the standard of our forefathers from the first settlement of the English colonies, and is exactly coeval with them. It is not unimportant that the standards, to be transmitted to the several states of the Union, should be of the same metal as the national standards, of which they shall be copies. The changes of the atmosphere produce different degrees of expansion and contraction upon different metals : and, when a measure of brass or copper is to be taken from a measure of platina, the differences of their expansibility become subjects of calculation, upon data not yet ascertained to entire perfection. The selection of platina for the French kilogramme has been attended with the singular consequence, that the standard of the archives is not of the same *weight* as the standard for use. The latter is of brass ; and the copies taken from it for the real purposes of life are of the same weight in the air, but not of the same weight as the platina standard, because that is the weight of the cubic decimetre of distilled water *in vacuo*. Whenever calculations of allowances for atmospheric changes in the different metals are introduced into the comparison of measures, estimates take the place of certainty ; and different results proceed from different times, places, or persons. The very immutability of platina, therefore, makes it unsuitable for a practical standard of mutable things. Change, and not stability, is the uniform measure of change. Justice consists in estimating every thing by the law of its nature: and, to illustrate this idea by applying it to moral relations, it may be observed, that, to bring mutable substances to the test of immutable standards, would be like charging disembodied spirits to pass sentence by the laws of their superior nature upon the frailties and infirmities of man.

The plan which is thus, in obedience to the injunction of both houses of Congress, submitted to their consideration, consists of two parts, the principles of which may be stated : 1. To fix the standard, with the partial uniformity of which it is susceptible, for the present, excluding all innovation. 2. To consult with foreign nations, for the future and ultimate establishment of universal and permanent uniformity. An apology is due to Congress for the length, as well as for the numerous imperfections, of this report. Embracing views, both theoretic and historical, essentially different from those which have generally prevailed upon the subject to which it relates, they are presented with the diffidence due from all individual dissent encountering the opinions of revered authority. The resolutions of both houses opened a field of inquiry so comprehensive in its compass, and so abundant in its details, that it has been, notwithstanding the lapse of time since the resolution of the Senate, as yet but very inadequately explored. It was not deemed justifiable to defer longer the answer to the calls of both houses, even if their conclusion from it should be the propriety rather of further inquiry than of immediate action. In freely avowing the hope that the exalted purpose, first conceived by France, may be improved, perfected, and ultimately adopted by the United States, and by all other nations, equal freedom has been indulged in pointing out the errors and imperfections of that system, which have attended its origin, progress, and present condition. The same liberty has been taken with the theory and history of the English system, with the further attempt to shew that the latter was, in its origin, a system of beauty, of symmetry, and of usefulness, little inferior to that of modern France.

The two parts of the plan submitted are presented distinctly from each other, to the end that either of them, should it separately obtain the concurrence of Congress, may be separately carried into execution. In relation to weights and measures throughout the Union, we possess already so near an approximation to uniformity of law, that little more is required of Congress for fixing the standard than to provide for the uniformity of fact, by procuring and distributing to the executives of the states and territories positive national standards conformable to the law. If there be one conclusion more clear than another, deducible from all the history of mankind, it is the danger of hasty and inconsiderate legislation upon weights and measures. From this conviction, the result of all inquiry is, that, while all the existing systems of metrology are very imperfect, and susceptible of improvements involving in no small degree the virtue and happiness of future ages; while the impression of this truth is profoundly and almost universally felt by the wise and the powerful of the most enlightened nations of the globe; while the spirit of improvement is operating with an ardor, perseverance, and zeal, honorable to the human character, it is yet certain, that, for the successful termination of all these labors, and the final accomplishment of the glorious object, permanent and universal uniformity, legislation is not alone competent. A concurrence of will is indispensable to give efficacy to the precepts of power. All trifling and partial attempts of change in our existing system, it is hoped, will be steadily discountenanced and rejected by Congress; not only as unworthy of the high and solemn importance of the subject, but as impracticable to the purpose of uniformity, and as inevitably tending to the reverse, to increased diversity, to inextricable confusion. *Uniformity* of weights and measures, permanent, universal uniformity, adapted to the nature of things, to the physical organization and to the moral improvement of man, would be a blessing of such transcendent magnitude, that, if there existed upon earth a combination of power and will, adequate to accomplish the result by the energy of a single act, the being who should exercise it would be among the greatest of benefactors of the human race. But this stage of human perfectibility is yet far remote. The glory of the first attempt belongs to

France. France first surveyed the subject of weights and measures in all its extent and all its compass. France first beheld it as involving the interests, the comforts, and the morals, of all nations and of all after ages. In forming her system, she acted as the representative of the whole human race, present and to come. She has established it by law within her own territories; and she has offered it as a benefaction to the acceptance of all other nations. That it is worthy of their acceptance, is believed to be beyond a question. But *opinion* is the queen of the world; and the final prevalence of this system beyond the boundaries of France's power must await the time when the example of its benefits, long and practically enjoyed, shall acquire that ascendency over the opinions of other nations which gives motion to the springs and direction to the wheels of power.

Respectfully submitted.

JOHN QUINCY ADAMS.

DEPARTMENT OF STATE, *February* 22, 1821.

APPENDIX.

A 1.

With a view to ascertain the existing varieties of fact in the weights and measures used at the several custom houses of the United States, and thereby the state of the standards in the several states, the following circular letter was, at the request of the Secretary of State, addressed, by the Register of the Treasury, to the collectors of the customs throughout the Union:

[CIRCULAR.]

TREASURY DEPARTMENT,
Register's Office, November 15, 1819.

SIR: I am requested by the Secretary of State to ask the favor of your early information to that department, relative to the standard of weights and measures, used at the custom house in the collection of the duties of the United States; and to observe, that it will be particularly acceptable to be informed, whether, in dry measure, any other than the Winchester bushel and its parts, is used; and the capacity thereof, that is, its diameter at the top and bottom, and its depth in inches, and tenths of inches. In liquid measure, whether wine or beer measures are respectively used for wines and beer, or, whether confined to wine measure, both for beer and wine. In respect to weights—whether the troy weight is at all used, and whether Dearborn's patent balance is altogether adhered to, in collecting the duties on articles paying duty by the pound or hundred weight.

I am, &c.

P. S. Be so good as to state the number of grains by your troy weight, which your avoirdupois pound weighs.

The following table is the result of the answers received from the collectors. As the cubic foot, or 1728 cubical inches of spring water, at the temperature of 56 degrees of the thermometer, weighs 1,000 ounces avoirdupois, the Winchester bushel of 2150.42 inches contains 77 lb. 12 oz. 7¼ drams of the same water, and the half bushel 38 lb. 14 oz. 3⅝ drams. The returns will shew how nearly the actual weights and measures correspond with those proportions. It will not be expected that experiments made at the custom house with scales adapted to heavy weights, in constant use, and with such water as was nearest at hand, should be marked with philosophical precision, or very minute accuracy. But it is evident they were generally made with great care, and, in several instances, repeated with various kinds of water. From the experiments of sir George Shuckburg Evelyn, it appears that the specific gravity of *distilled* water, at the temperature of 62, is 252½ grains troy to a cubical inch; and of that water the bushel should contain 77 lb. 9 oz. 1½ drams, and the half bushel 38 lb. 12 oz. 8¾ drams. Mr. Pollock found the specific gravity of the pump water at Boston to be 253.6042 grains troy to the cubic inch, and of that water the statute Winchester bushel should contain 77 lb. 14 oz. 8½ drams, and the half bushel 38 lb. 15 oz. 4¼ drams. From 38 lb. 12 oz. to 39 lb. may be, therefore, considered as the range within which the different kinds of fresh water should fill the correct standard copper or brass half bushel.

In several of the returns it is apparent that the weight certified includes that of the wooden vessel which held the water. By the experiment of the late venerable William Ellery, collector at Newport, it appears that of two half bushels of the same dimensions, one of copper and the other of wood, the former contained 1½ ounces more than the latter, and both of them half an ounce more of spring than of rain water: and by experiments of G. Davis, inspector at New Orleans, from whom a very interesting report was received, it appears, that, by weighing the wooden half bushel before it had been filled, and after it was emptied of Mississippi river water, there was a difference of nearly 15 ounces, to be accounted for partly by the absorption of the water into the wood, and partly by the adhesion of it to the sides and bottom of the vessel.

By the testimony of Mr. John Warner, a brass founder in London, much employed in making for country corporations brass standard weights and measures, duplicates of those in the exchequer, given before the committee of the house of commons in the year 1814, it was shewn that from the extreme difficulty and expense of turning a bushel measure truly cylindrical, the practice of the trade is, notwithstanding the act of parliament, to pay no attention to the dimensions of the measures which they make, but to rely entirely upon the trials by the weight of water which they contain. From all the admeasurements which have been made of the English standards, it is apparent that neither the weight of water which a bushel or half bushel may be found to hold, nor the direct measurement by the depth and diameters of the vessel, can be relied upon by itself to ascertain its exact capacity; and even when one of these tests is applied as a check upon the other, the result may differ to the extent of four or five inches upon a half bushel. The corn gallons at the exchequer, which in 1688 were found by a skilful artist to be of 272 cubic inches, in 1758 were, when remeasured by order of the committee of the house of commons, returned as of 271 or less, and one of them tried in April, 1819, by sir George Clerk and Dr. Wollaston, appeared by the weight of water which it held, to be only of 270.4 inches.

If this apparent diminution should be attributed to the decay of the vessel in the lapse of time, that will not account for the opposite result of the two experiments upon the bushel, which, by direct measurement in 1758, was found to be of 2124 cubical inches, and in 1819, by the weight of water it contained, was of 2128.9 inches. The statute Winchester bushel is a cylinder 18½ inches in diameter and eight inches deep: a difference of $\frac{1}{100}$ of an inch either in the depth or diameter increases or diminishes the contents of the vessel nearly three cubical inches. Mathematical instruments are now constructed by which the division of $\frac{1}{10000}$ part of an inch may be discerned; but such refinement of art cannot be applied to the making of vessels of the size of a bushel, or its half. In all such vessels inequalities in the depth, diameter, or circumference, are unavoidable, producing differences in their capacity of more than five cubical inches; the test, therefore, by the weight of water they will hold, is more effectual for accuracy than that of measurement; its results, however, depend upon the correctness of the scales and weights as well as upon the care and attention with which the experiment is made.

As the patent scales of Dearborn are used in most of the custom houses throughout the Union, few of them possess the avoirdupois heavy weights. The correctness and convenience of the patent scales are generally attested by the collectors, who have them in use, and may be relied upon with all the confidence of which any steelyard can be susceptible.

The beer measure and the troy weight are seldom used except in the principal and most populous ports. Of forty single avoirdupois pounds the average weight in troy grains was 6998. In all the principal ports they were exact, within one grain, with the exception of New York, where the custom house pound was five or six grains over weight: but where that of the city was exact.

NAME OF THE COLLECTOR FROM WHOM THE RETURN WAS RECEIVED.	STATE AND PORT		WEIGHT OF WATER IT CONTAINS, AVOIRDUPOIS.			GRAINS TROY IN THE lb AVOIRDUPOIS.	REMARKS
			lbs	oz.	dr.		
...ngate, L. jun.	Bath		37	1		spring	7,039	Winchester measure.
...oe, D	Belfast					do.		do no troy weights ditto
...vis, Edward S	Frenchman's Bay		42	3	12*	ret		do do "no troy weight Winchester intended weight of the lb.* bushel measure
								*This mss include the
...rer, Joseph	Kennebunk		39			rain	7,039	Dearborn at Kennebunk Scales and weights at Wells. Standard from the town of Arundel.
...ith, George S.	Machias		37	10		rain	6,996	Dearborn. Troy lb no troy weight Winchester.
...acley, Stephen	Lubeck		38	8		spring	6,998	Dearborn. The standards are kept at East port.
							7,000	Winchester.
...k, Josiah	Portland and Falmouth		39	7		spring	6,998	Des la n's scale used
...cy, Isaac			38	3		spring	7,004	Dearborn principally before the late war
...unger, D	Saco		38	7	8	rain	7,035	used. No troy
...tk, Francis	Wiscasset		40			rain	7,032	Dearborn
			39			spring	7,013	or thereabouts, D.
...nam, Timothy	Portsmouth,	½ bushel	78			spring	7,004	Winchester. These, except the pound of 7,004 grains, are the standard
	New Hampshire.							Hardened copper measures and weights of the
	Portsmouth, ¼ bushel		38	12		do	4 lb. 27,996	state of New-Hampshire, imported from England before the Revolution, and dep.
								house ed, by Mr. Upham's request, at the cu
...rborn, Henry	Boston, copper		39	2		rain	7,000	Dearborn usd
...rquand, Joseph	Newburyport		39	4		rain	6,938	do. do. Typ not us
...tredge, John	Gloucester		38			spring	6,999	do. by an oun of 437
...lps Nathaniel	Dighton, copper		37	6		rain	6.99	
...vys John	New Bedford		38	14	12	rain	7,000	Dearborn
	4 bushel tub							
...en, Isaiah L.	Barnstable		38	8		spring	6,999	do. No troy
...k, Thomas, Jr.	Edgartown		38			do	6,999	do.
...ton, Martin T.	Nantucket		38	4		rain	6,978	do.
	Rhode Island							
...s, Thomas	Providence, copper		39			rain	6,87	do No troy
...ng Charles	Bristol, bushel		78	14	12	rain	6,984	do do
...ly, William	Newport, copper		38	15	4	spring	6,999	
			38	15		spring		
	Connecticut.		38	13	12	spring		Winchester. Dearborn's patent not used for hemp and iron.
...ing, T. H.	New London	wooden	38	13		rain	6,995	Winchester. Dearborn's patent not used. A state law, of the revision of January, 1798 prescribes gauging by Gunter's scale, and a rule to find the mean dep. of casks.
...dley, Walter	Fairfield		39	8		spring	6,969	Dearborn
								Th law of Con make no mention of the Winchester bushel.
...ton, David	New York. copper		78	13		rain	7,007	by N. Y. Bank scales Th New York city jik too short, being

...E OF THE COLLEC- ... FROM WHOM THE ... URN WAS RECEIV-	STATE AND PORT.	Top	Bottom.					GRAINS TROY IN THE POUND AVOIRDUPOIS.	REMARKS
		Inch. dec	Inch. dec.	lb. oz. dr					
	New York.								
wley, I. -	Rochester, a ½ bushel	13.5	14	-				Sealer's own 4?? of which the pound is 7,008	} Dearborn's balance used The three half bushels belonged to diffe merchants, they and the sealer's weights and measures w?? from Albany.
	and	13 5	14						
	another	13⅝	13⅞						
	New Jersey.								
;wster, John -	Perth Amboy -	Wi??	r bushel sea					6,9?? - -	Beam and scale with avoirdupois weight only used.
	Pennsylvania.								
.ele, John -	Philadelphia, copp r	13.8	12	39 6 Schuylkill				700 - -	Dearborn's balance not used.
	Delaware.								
Lane -	Wilmington -	12.5	11.5	38 10 6 spring				7,00 - -	D?? ı t?? weight not u??
	Maryland.								
Culloch, Jas H.	Baltimore, brass	18.5	18.5	77 8 rain				7,000 - - -	Dearborn. Th wine gallon contains 8lb. 2oz. a?? rain w??
.llis, John -	Oxford, two ½ bush.	12 7	12	40* spring				No troy but for m?? -	*Weight of the ?ess ls must be included G?? d wheat weighs by th?? fr?? 58 to 62 p?? per bus el
	District of Columbia.								
	Washington† -	13	13	38 3 13 mixed				7,01 No troy -	† See the note on the standards of the D?? of Columbia
	Georgetown, peck	10.5	9.5	19 7 6⅔				6,97	
ıson, Thomson	Alexandria -	14.47	14.47	38 13 8 river				6,950 - -	Dearborn's bala used, except for ıro
	Virginia.								
ll?d Nathaniel	Cherrystone -	14 2	13.1	41 10* - -				- -	*M?? include the weight of the vessel.
.llory, Charles K.	Norfolk, copper	18 4	18.4	78 r?? salt				- - -	D earborn, ex?t fo salt
??s d s ph -	Petersburg -	13.5	13.5	39 river				6,999½ - -	Dinwiddie co?n standards.
.bon, I -	Ric??d copper	18⅝	18⅝	77 8 w??				6,984 - -	H nrico co?n standards.
	North Carolina.								
wyer, ?En ch -	Camden -	13.7	13.7	39 12 spring				Troy not used -	D?? ı ?alan e used altogether.
edwell, Samuel	Edenton -	14	12.9	38 11 rain				701 4 do. not us?d -	The weights wer procured f?? N?? York.
wks, Fran??	Newbern -	13 9	13.4	43 12* spring				7,032 not u?? -	Standards ?? from England *M?? include the w?? of the half bushel
ıgleton, ?? s S.	Ocracoke -	13 75	13.75	38 w?? or spring				do. do. -	*Dry measure standard from N?? Y?? :*
gan, L -	Plymouth -	13 7	13.7	38 8 r??				No troy weights -	?? ı must be an error in this measure.
?, Thomas H	W?s ıngton -	13⅞	14$\frac{1}{16}$	36 6 spring				6,?? - . - -	D arborn's balance used an? preferred
	South Carolina.								
ngle, Ja?? R.	Charl?? -	16.8	16.1	78 saltish well -	23			6,9?? Troy not used -	Dearborn's balance u?? altogether
				77 9 8 cistern not clea					
	Georgia.								
.llock, A. S. -	Savannah -	14	12.7	38 r??		Wine	both	6,999½ do. -	ditto.
ırk, Frederick	St. Mary's -	13 25	13.25	38 12 spring		Wine		- - - -	dito
				39 8 rain					
	Louisiana.								
ew Beverly -	New Orleans -	14.13	13.79	39 6 2 13 dry	52	Wine	Bee	6,984 - - -	Dearborn's balance used. A dozen bott of wine or of beer are found to co?? 2½ gallons of each, according to the?? resp t-ıve sta??
				38 7 5 we, ıer				6,?9	
vıs G. ??	2 bushel tub	17.8966	13.75					7,?0 by 2 lb. of 14.008	
	Alabama.								
wıs, Ad? -	Mobile - -	The measu	res were obt			W?e	B er	Dearborn's ba?? -	Th Mississippı co?n sta?? were obtained f?? Philadelphia.

APPENDIX.

A 3.

Note on the weights and measures of capacity in the District of Columbia.

By the constitutional law of the District, the standard weights and measures of Alexandria are derived from Virginia, while those of Washington city and Georgetown are from the standard of Maryland. The law, in both states, as has been shewn, is, and for more than one hundred and fifty years has been, the same, namely : the act of parliament of 1496, for the statute book. The Winchester measure, avoirdupois weights, and the Exchequer standards for practice. The avoirdupois pound of Alexandria is one of the most defective in the Union, being 50 grains troy, or more than two pennyweights, short of 7,000 grains. The half bushel is also too small ; its primitive standard having been, not the Winchester bushel, but one of the oldest Exchequer standards, made under the statute of 1266. It is 16 cubical inches less than the parliamentary Winchester half bushel.

The Georgetown peck is in exact proportion to the Winchester bushel. Its pound avoirdupois is too light by 30 troy grains.

The new measures at Washington lately obtained at Baltimore, are a half bushel, peck, and half peck, of copper, lined with tin, cylindrical in form ; the half bushel and peck with a hole and cock in the centre of the bottom, to let off the water.

The half bushel appears to have been intended to be of 13 inches diameter, and eight inches depth, which would have given 1064 inches of cubical contents. This would correspond almost exactly with *one* standard bushel at the English Exchequer, which was tried in April, 1819, by weight of water, and found to hold 2128.9 inches ; but, of the new Washington half bushel, neither the depth nor the diameter are uniform. They vary to the extent of a quarter of an inch, in different parts of the vessel. The bottom is not perfectly flat, but bulges a little inward at the centre ; nor is the edge of the top circumference perfectly level ; so that, while on one side it overflows, it is left not entirely full on the other. Its mean diameter is, with sufficient exactness, 13 inches, and its depth 7.9, which gives 1058.6 inches for its cubical contents. This is 16.61 inches less than it ought to be.

It was found to contain 46lb. 5oz.13 pwt. or 267,672 grains, troy weight, of mixed rain, river, and pump water, with Fahrenheit's thermometer at 42°, and then overflowed at one side of the border, while on the opposite side there was about $\frac{1}{10}$ of an inch yet to fill. This also gives 1058.6 inches as its capacity. It contained of wheat, in stricken measure, 31lb. 4oz. 5½dr. avoirdupois, of which the old half bushel had contained 32lb.

The peck has in its dimensions the same irregularities, though in less degree than the half bushel. Its diameter is ten inches ; its depth from 6.7 to 7, with a bottom bulging inward. The mean depth is about 6.9, and its contents, by that measure, are 541 inches.

APPENDIX.

It holds 23lb. 8oz. 15 pwt. or 136,680 grains troy of water, giving also 541 inches, being 3.4 inches more than the standard measure.

The half peck is 7.5 diameter, and 6.25 inches deep, which gives 276.25 inches for its contents. It holds 12lb. 1oz. 4pwt. or 69,696 grains troy of water, or 276.5 inches, being 8 inches more than the standard measure.

Although each of these measures, separately taken, is incorrect, they are so far just in the aggregate, that the half bushel and the peck, with the half peck twice filled, would yield, with perfect exactness, a bushel of the legal standard of 2150.42 cubic inches.

The half bushel, peck, and half peck, which had been used as standards at Washington until the last autumn, were much more exact, both in their proportions to one another, and in their conformity to the Winchester bushel, than those now substituted in their stead. The one pound avoirdupois had been of 6,962 grains, or 38 grains too light. The weights now are

1 pound	7,010 grains troy	10	
2	14,024	24	} grains too much.
4	28,062	62	

The wine gallon is of 231 solid inches, and the smaller measures proportional to it. There are no standards either of troy weight or beer measure.

	WINE GALLON		CORN AND BEER GALLON.	
	Statute. Cubic inches.	Existing Standard. Cubic inches.	Statute. Cubic inches.	Existing standard. Cubic inches.
1225 By Magna Charta, c. 25, explained by 1266 St. 51 Henry III, and by the treatise of 1304 weights and measures of 31 Edward I. †The pound sterling of 12 ounces = 5,400 grains troy and pound for other things of 15 ounces = 6,750 grains troy	-	-	-	-
1353. By assize of the tun. Statute of provisors, 27 Edward III. stat. 1, ch. 8. 1459.‡ And by St. 18 Henry VI. ch. 17	217.6	217.6 Irish gallon	266.3	Rumford gallon of 1228 2130.4 2134 2128.9 Exchequer do.,
1496.6 Statute of 12 Henry VII. ch. 5; according to its letter with troy pound of 12 ounces	219.43	217.6 do.	268.53	do. 2148.24 2150.42 do.
1496. Same statute, with application of the rule *Compositio Mensurarum* of 1266, to the pound troy of 12 and of 15 ounces	224	224 Guildhall gal. in 1688	224	None 2145.6 Winchester bushel a do. 1792
1496 Same statute, with application of the rule of 1266 to the pound of 12 ounces	224	224 do.	280	Exchequer bushel of Henry VII. 2294 2240 Exchequer pint of 1601 and 1602 2217.62 Coal bushel by act 12 Anne,
1532. Statute 24 Henry VIII. ch. 3.	-	-	272	272 Exchequer quart of 1601 271 Gallon of 1601 marked E. E. 2176 None.
1495 Same 12 Henry VII. ch. 5, with application of the rule of 1266, and wheat of 32 lbs. to 22½ grains troy, and bushel as it prescribes was executed. Then in such stand.	224	do.	-	270.4 Gallon of 1601 marked E.
1531 ‖ Stat. 23 Henry VIII. ch. 4, sec. 13. 1705. Stat. 5 Anne, ch. 27, sec. 17.	281	None before 5 Anne	282	282 Treasury ale gallon 2256

* The object of this table is to show the varieties of the ancient standard measures of capacity in England correspond very exactly with the variations of

1st. That the statute of 1256 the bushel was of capacity to contain wheat equal not to the measure, but to the *weight* of eight gallons of wine.

2d. That by the statute of 1496, conformably to its professed intention, not only changed the pound sterling for the pound troy, but, by its letter, changed the capacity of the bushel from the weight of eight gallons of wine, to the measure of eight gallons troy weight of wheat, which was a gallon of 224 cubic inches.

3d. That, as far as concerns the corn or ale gallon, and the bushel, thus act of 1496 was never executed. The ancient stand- and bushel, so far as it prescribes was executed.

4th. That by the statute of 23d Henry VIII. chapter 4, section 13, the ale gallon is declared equal to one-eighth of the measure of the bushel, with express reference to the Compositio Mensurarum of 1266

1266   *Statute 51st Henry III Compositio Mensurarum and treatise of* 1304.

1	Penny Sterling round and unclipped shall weigh	32 Wheat Corns.
1	Ounce	20 Pence.
1	Pound	12 Ounces.
8	Pounds	1 Gallon of Wine.
8	Gallons of Wine	1 Bushel.

† *Statute 18th Henry VI. chapter* 17. *Ancient assize of the Tun.*

Measure of one Bushel   1 Tun = 252 Gallons; 1 Gallon = ⅛ of 8 Cubic Feet, = 219.43 Cubic Inches.
1 Hp   126
1 Tertian 84
1 Hogshead 63

Statute 12*th Henry VII. chapter* 5, 1496.

Measure of one Bushel
Every Gallon   8 Gallons of Wheat.
Every Pound   8 Pounds Troy Weight of Wheat.
Every Ounce   12 Ounces Troy Weight.
Every Sterling   20 Sterlings, [meaning Pennyweights Troy
                 32 Corns 1 With

‖ *Statute* 23*d Henry VIII chapter* 4, *section* 13, 1531

Every Ale Barrel shall be made according to the assize specified in the rule of 8 gallons of the said assize of the rule bushel, make the corn bushel.

APPENDIX. 143

C.

Note, on the proportional value of the pound sterling and the dollar.

The whole amount of the commercial intercourse between two countries, within a given time, say a year, may be considered as the barter of an equivalent portion of their respective productions. The balance of trade is the excess of exportation from the one, and of importation to the other, beyond the equivalent value of specific articles of the trade.

In the practice of commerce, all the articles of the trade are valued in the established currencies of both countries; each article first in the country from which it is exported, and secondly in that to which it is imported. The balance of the trade must be discharged by some article of equal agreed value to both parties. There are two precious metals *gold* and *silver*, which, by the common consent of all commercial nations, are such articles; and there is no other.

These two metals constitute also the principal basis of the money, or specie currency of all commercial countries; and as they are variously modified by weight and purity in the *coins* of different countries, a common standard must be resorted to, by which the relative value of the coins of the two countries may be ascertained and settled in their commercial dealings with each other.

Some one specific coin or money of account on each side is assumed between which a proportional value is established as the conventional par of exchange. Thus, between the United States and Great Britain, the dollar of the former and the pound sterling of the latter, with their respective subdivisions, are assumed as the standards of comparative value, and the conventional proportion of value between them, commonly used in their commercial intercourse, and sanctioned by several acts of Congress, has settled the *par of exchange* at one pound sterling for four dollars and forty-four cents in the United States, while, in Great Britain, it is at four shillings and six pence for the dollar.

But observe:

First, That here are already two different bases of exchange—the American, which assumes the pound sterling for the unit, and estimates it in the proportional parts of the dollar, and the English, which assumes the dollar for the unit, and values it in the proportional parts of the pound sterling. This would have been immaterial, if the calculations upon which the exchange was originally settled, had been correct. But the results of the two estimates are not the same. If the dollar is worth four shillings and six pence, the pound sterling is equivalent to four dollars forty-four cents four milles, and an endless fraction of four decimal parts. If the pound sterling is worth four dollars and forty-four cents, four shillings and six pence, or fifty-four pence, are equal only to ninety-nine cents and nine milles. The difference is of one mille in a dollar, or one thousand dollars in a million.

Secondly, That the elements of this exchange, the two objects of comparative estimated value, are not homogeneous. The dollar of the United States is at once a money of account, and a specific silver coin, while the pound sterling, at the time when the exchange was settled, was only a money of account, having no coined representative in one piece of either of the precious metals. Since that time, indeed, the pound sterling has found a spurious representative in paper notes of the Bank of England, and of late a more truly sterling representative in the piece of gold which is called a *sovereign*. So that the pound sterling in England is an indefinite term, represented by three different materials, that is, in gold, by the sovereign, or by the guinea, with deduction of a shilling; in silver, by twenty shillings, or four crowns, or in paper, by a bank note.

In the United States, their coins, both of gold and silver, are legal tenders for payments to any amount; but in England silver coin is a legal tender for payments only to an amount not exceeding forty shillings, and by the restrictions of cash payments by the bank, the only actual currency, the only material in which an American merchant having a debt due to him in England can obtain payment is Bank of England paper. So that at this time the materials of *exchange* between the United States and England are, on the side of the United States gold or silver, on the side of Great Britain, bank paper.

Suppose an American merchant has a debt due to him in England, which is remitted to him in gold bullion, or coins of the English standard, say £10,000. He receives of pure gold 196 pounds, 2 ounces, 3 pennyweights, 22 grains, for which, when coined at the mint of the United states, he receives 45,657 dollars 20 cents. The pound sterling, therefore, yields him 4 dollars 56.572 cents. And such is the value of the pound sterling, if the par of exchange be estimated in gold, according to the standard of purity common to both countries.

If the payment should be made in silver bullion, at 66 shillings the pound troy weight, according to the present English standard of silver coinage, he would receive only 43,489 dollars and 43 cents, and the pound sterling would only nett him 4 dollars 34.8943 cents.

The pound sterling, therefore, estimated in gold,
is worth - - - - - $4 56.5720
In silver 4 34.8943

Making a difference of 21.6777
Half of which 10.8388

Added to $4 34.8943
And deducted from 4 56.5720
Makes what is called the medium par of exchange, $4 45.7331.

It is contended by some writers upon the commercial branch of political economy, that this medium is the only equitable par of exchange; but this is believed to be an error. It is, perhaps, of as little importance what the conventional par of exchange is, as whether a piece of linen or of broadcloth should be measured by a yard or an ell. The actual exchange is never regulated by the medium or any other par, but by the relative value of bullion in the two countries at the time of the transaction ; by the relative proportions between the value of gold and silver established in their respective laws ; by the prohibitions of exportation of bullion sometimes existing, and the duties upon its exportation levied at others ; by the laws, which, in some countries, make gold alone, in others silver alone, in others again both silver and gold, legal tenders for the payment of debts ; by the existing condition of the commerce of the two countries, and of each of them with all the rest of the world ; and last, and most of all, by the substitution of paper currency instead of the precious metals, in one or both of the countries, and the existing depreciation of the paper.

But the law of the United States, first enacted on the 31st July, 1789, sect. 18, prescribing that, for the payment of duties, the pound sterling of Great Britain shall be estimated at 4 dollars 44 cents, [U. S. Laws, Bioren's edition, vol. 2, page 22] is not so indifferent. This provision of the law has been continued in both the collection laws, since enacted, and, by that of 2d March 1799, [3 U. S. Laws, sect. 61, page 193] is still in force.

By section 30 of the act of Congress of 31st July, 1789, the duties were made receivable in gold and silver coin *only ;* the gold coins of France, England, Spain, and Portugal, and all other gold of equal fineness, at 89 cents per pennyweight ; the Mexican dollar at 100 cents ; the crowns of France and England at one dollar and 11 cents each, and all silver coins of equal fineness at one dollar and eleven cents per ounce.

As this was one of the first experiments of legislation under the present constitution of the United States, it is unnecessary to make upon it many of the remarks which suggest themselves ; but, with regard to those of its provisions which are still in force, let us observe,

That, on the 31st July, 1789, there had been no suspension of specie payments by the Bank of England. The pound sterling, if paid in gold, yielded 113.0014 grains of pure metal. If paid in silver, 1718.72 grains of pure silver.

That the dollars and cents in which this pound sterling was estimated by the act of 31st July, 1789, were not the dollars and cents of the standard now established, but of the standard established by the resolution of the old Congress of 8th August, 1786, and their ordinance of 16th October of the same year, [1 U. S. Laws, page 646] by which the dollar was to contain 375.64 grains of pure silver, and the eagle 246.268 grains of pure gold.

This dollar had been assumed as the money unit of the United States, upon a report from the Board of Treasury, dated 8th April, 1786, from which report it appears that the board intended and believed that it would be of equal value with the Spanish dollar, then generally current in the United States at four shillings and six pence sterling; excepting an allowance which they proposed to make for the waste and expense of coinage of silver. They made a similar allowance of ½ per cent. upon the coinage of gold.

The ordinance assumed, for the standard of purity, both of gold and silver coins, eleven parts fine, and one part alloy. This standard was, with respect to gold, the same as that of England. But the English standard of silver coins is eleven ounces and two pennyweights of fine, to eighteen pennyweights of alloy ; so that, while the English pound troy weight, of coined silver, contained 5,328 grains of pure metal, that of the United States, by the standard then established, contained only 5,280 grains.

In the elaborate calculations of the report, which were adopted as the basis of the ordinance, no allowance whatever is made for this difference of 48 grains in the pound troy, between the English standard and that prescribed for the United States. It expressly states that the English mint price of standard silver is sixty-two shillings sterling, and professes to prepare a dollar of *equal* value, excepting an allowance of two per cent. for waste and coinage. It then draws a proportion without reference to the difference between the two standards, and computes the sixty-two shillings of the English standard pound troy, as if they contained only 5,280, while they really contained 5,328 grains. The object of this omission apparently was, together with the two per cent allowance for waste and coinage, to preserve what the report states to have been the proportional value established by custom in the United States, between coined gold and silver of fifteen and six tenths for one, while their proportional value in the English coins was 15.21 for one.

The ordinance for the establishment of the mint, and for regulating the value and alloy of coin, therefore prescribed that bullion, or foreign coin, should be received there as follows :

Uncoined gold, or foreign gold coin, eleven parts fine, and one part alloy - - - 1 lb. troy weight $209 77
Silver, 11 parts fine and 1 part alloy 1 lb. - - 13 77 7

And so in proportion to the fine gold and silver in any other foreign coin, or bullion. And the dollar to be issued from the mint of the United States was settled at 375.64 grains of pure silver, because the report of the Board of Treasury had first supposed, contrary to that fact, that there were only 5,280 grains of pure silver in sixty-two shillings of English silver coin, consequently, only 383.225 grains, instead of 387, in four shillings and six pence, and then provided an allowance of two per cent. for waste and coinage. By these operations it seems to have been thought that the standard dollar of the United States would be of equal value with the Spanish dollar, then current in this country, and with four shillings and sixpence of English silver coin. Thus, while, by the 18th section of the act of 31st July, 1789, the pound sterling was estimated, for the payment of duties, at four dollars and forty-four cents, by the 30th sec-

APPENDIX.

tion of the same act, every pound sterling paid in guineas, or other gold, was received for $4.57.143, and if paid in English crowns, was received for $4.57.5445.

That the calculations upon which the rated value of gold and silver coins was fixed were loose and inaccurate, is apparent. The gold coins of France and Spain were rated as of the same standard of purity with those of England and Portugal; the crown of France as of equal value with the English crown; both without reference to their weight, and both as equivalent to an ounce of silver of the same fineness. It was well known and intended that all these coins should be rated at more than their intrinsic value, compared with the pound sterling, as estimated at 4 dollars 44 cents, or with the standards of gold and silver coins of the United States then established. The differences might be considered in the nature of a discount for prompt payment of the duties. And, as the merchants of the United States were deeply indebted in England, inasmuch as the pound sterling was undervalued, the difference was clear profit to them in discharging the balances due to their English creditors.

The act of 31st July, 1789, was, at the succeeding session of Congress, repealed, and that of 4th August, 1790, substituted in its stead, [2 U. S. Laws, p. 131.] The 40th and 56th sections of this act correspond with the 18th and 30th sections of that of 1789. The pound sterling is again rated at $4.44, and the coins as before

But on the 2d April, 1792, passed the act establishing a mint and regulating the coins of the United States : by which the whole system established by the ordinance of 1786 was abandoned, and different principles and different standards were assumed. The standard of gold coins was left at 11 parts fine to one of alloy; but instead of 246.268 grains of pure gold, the eagle was required to contain $247\frac{1}{2}$ grains. The silver standard was altered from 11 parts in 12 of fine, to 1485 parts in 1664. Instead of 375.64 grains of pure silver, the dollar was required to contain only $371\frac{4}{16}$ grains, and its weight, instead of 409 grains, was fixed at 416. The proportional value between gold and silver was fixed by the same law, at fifteen for one; and instead of the allowance of two per cent. for waste and coinage, the principle was adopted of placing gold and silver coined at the same rate as uncoined, and of delivering at the mint coined, the same weight of pure metal as should be brought to it in bullion or foreign coin.

By this operation the value of the silver dollar as compared with British silver coin was reduced from 52.4539 pence sterling to 51.8409 pence; and the pound sterling, from $ 4.57.5445, was raised to be worth $ 4.62.955; and, at the same time, the value of the dollar estimated in the English gold coin was raised from 52.304 to 52.5656 pence, and the pound sterling was reduced in the gold coin of the United States from $4.57.143 to $ 4.56.572.

The act establishing the mint had, however, no direct reference to the value or the rates of foreign coins. But on the 9th February, 1793, passed the act regulating foreign coins, and for other purposes,

[2 U. S. Laws, p. 328,] which made the gold coins of Great Britain and Portugal of their then standard a legal tender for the payment of all debts and demands at the rate of 100 cents for every 27 grains of their actual weight. The gold coins of France and Spain at the rate of 100 cents for every $27\frac{2}{5}$ grains : Spanish dollars weighing not less than 415 grains at 100 cents : French crowns weighing not less than 459 grains, 110 cents each. The 55th [56th] section of the act of August, 1790, was repealed, but the 40th section was left in force, and the pound sterling was still receivable for $4.44. It was, however, thenceforward, whether paid in the gold coins of England or of the United States, worth $4.56.572.

A new collection law was enacted on the 2d March, 1799, which is still in force. In the 61st section of which [3 U. S. Laws, p. 193,] the pound sterling of Great Britain is again rated at 4 dollars 44 cents; while, in the 74th section, the gold coins of Great Britain of the standard prior to 1792, are receivable at the rate of 100 cents for every 27 grains. But a proviso is added to the 61st section, that the President may establish regulations for estimating duties on goods, invoiced in a depreciated currency; and a proviso to the 74th, that no foreign coins but such as are a lawful tender, or made receivable by proclamation of the President, shall be received.

In the act of 9th February, 1793, the English crown was not rated at all, and from that time no English silver coin has been a legal tender, nor consequently receivable in payment of duties.

The act of 10th April, 1806, regulating the currency of foreign coins in the United States, continued the rates established by the 74th section of the act of 2d March, 1799; and it required of the Secretary of the Treasury to cause assays to be made every year, and report them to Congress, of the foreign coins made tenders by law, and circulating in the United States.

29th April, 1816, [6 U. S. Laws, p. 117.] Act regulating the currency within the United States of the gold coins of Great Britain, France, Portugal and Spain, the crowns of France, and five franc pieces.

Gold coins of Great Britain and Portugal 27 grs. = 100 cts. or $88\frac{8}{9}$ cts. per dwt.

France	$27\frac{1}{2}$	= do.	$87\frac{1}{4}$
Spain	$28\frac{1}{2}$	=	84

Crowns of France, weighing 449 grs. 110 cents, or $1.17 per oz.
Five franc pieces 386 grs. 93.3 1.16 do.

3d March, 1819. Act to continue in force the above act.

After 1st November, 1819, foreign gold coins cease to be a tender. Rest of the act to be in force till 29th April, 1821.

The act of 2d April, 1792, establishing the mint, was founded, in its principal features, upon the report of the Secretary of the Treasury, Hamilton. It is remarkable that in this report all notice of the ordinance of Congress of 16th October, 1786, is omitted.

It says, " a prerequisite to determining with propriety what ought " to be the money unit of the United States, is to endeavor to form

APPENDIX. 119

"as accurate an idea as the nature of the case will admit of, what it actually is. The pound, though of various value, is the unit of the money of account of all the states. But it is not equally easy to pronounce what is to be considered as the unit in the coins. *There being no formal regulation on the point* (the resolutions of Congress of the 6th July, 1785, and 8th August, 1786, having never yet been carried into operation) it can only be inferred from usage or practice."

Now the ordinance of 16th October, 1786, was a formal regulation, which recognized the principles, in regard to the unit of coins, of the resolutions of 6th July, 1785, and 8th August, 1786; and the Congress, under the new constitution, had, by the two successive collection laws of 31st July, 1789, and 4th August, 1790, not only rated the foreign moneys of account, but foreign coins, by the standard of dollars and cents, established in the resolution of 8th August, 1786. Millions of dollars had been received in revenue, under those laws, in foreign coins estimated in those dollars and cents. A pamphlet was published by Mr. Boardley at Philadelphia, in 1789: in which he shews that the real value of the dollar, in the first collection law, was 52.46 pence sterling, and not 54, and adds: "I do not consider whether this valuation accords with a late declaration that twenty shillings sterling shall be estimated at the value of 4 dollars and 44 cents of the present dollar; but I recommend it to the consideration of others."

In the Gazette of the United States, of 24th October, 1789, is an essay entitled "A few thoughts concerning a proper money of account, by a gentleman of Virginia," in which it is fully shewn that the valuation of the pound sterling, "as it stands rated by Congress at 4 dollars 44 cents," was inconsistent with the pennyweight of gold rated at 89 cents; that the pound sterling should be rated at 4 dollars $57\frac{89}{623}$ cents, or the pennyweight of foreign gold coin at $86\frac{19}{36}$ cents, instead of 89, which it states to be greatly to the injury of the revenue.

The alterations from the system established by the old Congress, recommended in Mr. Hamilton's report, and adopted by the law for establishing the mint, were, a dollar of $371\frac{1}{4}$ grains pure silver, instead of 375.64 grains; an eagle of $247\frac{1}{2}$ grains pure gold, instead of 246.268; 15 for 1 proportional value of silver and gold, instead of 15.6 for 1. Gratuitous coinage, instead of a duty of two per cent. for the bullion sent to the mint to be coined.

Mr. Hamilton proposed to leave the standard of purity of the silver coin at 11 parts in 12 pure, as it had been established by the old Congress. But, in this respect, the law departed from the principles of the Secretary. It took the weight as well as the pure contents of the Spanish dollar, then in circulation, for a model; not indeed its legal weight and purity, which would have been 420 grains, at $10\frac{3}{4}$ parts in 12 pure silver; but its actual weight and purity, with the allowances for remedy, and ascertained by the average from a considerable number of the Spanish dollars, of the coinage since 1772, which were then in actual circulation. The result gave us a dollar of 416 grains, and containing 371¼ grains of pure silver.

In the coins of the United States there is no allowance for what is called the remedy of weight; but assays of all coins issued from the mint are made, and if any of them are found inferior to the standard prescribed, to the amount of more than $\frac{1}{144}$ part, the officers of the mint, by whose fault the deficiency has arisen, are to be dismissed. This provision was adopted from what was stated in Mr. Hamilton's report to be the practice of the mint in England.

By the acts of incorporation of the Banks of the United States, their bills, *payable on demand*, are made receivable in all payments to the United States, unless otherwise directed by Congress.

By the acts of 31st July, 1789, and 4th August, 1790, the gold coins of Great Britain were rated at 89 cents the pennyweight. By the act of 9th February, 1793, passed after the change of the standard of our domestic coins, British gold coins were rated at 27 grains to the dollar, equivalent to 88$\frac{8}{9}$ cents the pennyweight, at which they stand to this day.

In the year 1797 the British parliament passed an act restricting the Bank of England from paying their own notes in specie, a restriction which has been continued to this day, with certain exceptions, by recent acts of parliament. The pound sterling, therefore, in all English invoices and accounts, is now neither gold nor silver, but bank paper. This paper has been at times so depreciated that Spanish dollars have been issued by the bank itself, successively, at five shillings and five shillings and sixpence the dollar, and they have passed in common circulation at six shillings.

In the year 1816 there was a coinage of silver at the mint, in which the pound troy weight of standard silver was coined into 66 shillings, instead of 62 shillings, which had been the standard before.

And an act of parliament of 2d July, 1819, confirms the restrictions upon cash payments by the bank, until the first day of May, 1823, with the following exceptions.

1. That, between 1st February and 1st October, 1820, any person tendering to the bank its notes payable on demand, to an amount not less than the price or value of sixty ounces of gold, at the rate of four pounds one shilling per ounce, shall receive payment in gold, of the lawful standard at that rate of £4 1s. per ounce.

2. That, from 1st October, 1820 to 1st May, 1821, such payments shall be made in gold, calculated after the rate of £3 19s. 6d. per ounce.

3. And that, from the 1st of May, 1821, to the 1st of May, 1823, they shall be make in gold, calculated after the rate of £3 17s. 10½d. per ounce. All these payments to be made, at the option of the bank, in ingots or bars, of the weight of sixty ounces each, and not otherwise.

Throughout this whole canto of mutability, the pound sterling of Great Britain, from the 31st July, 1789, to this day, has been rated by the laws of the United States, at 4 dollars and 44 cents.

APPENDIX.

There has probably been no time since the establishment of the mint of the United States, nor since the first establishment of the dollar as the unit of account in the moneys of the United States, when this has been the intrinsic value of the pound sterling, whether computed in gold, silver, or bank paper.

A proclamation of Queen Anne, issued in the year 1704, declared that the Spanish, Seville, and Mexican, *pieces of eight*, (as dollars were then called) had, upon assays made at the mint, been found to weigh seventeen pennyweights and a half (420 grains,) and to be of the value of four shillings and sixpence sterling, from which the inference is conclusive that they contained of pure silver 387 grains, and the proclamation accordingly prohibited their passing, or being received, for more than *six* shillings each, in the currency of any of the British colonies or plantations. An act of parliament in 1707, corroborated by penalties the prohibition contained in the proclamation. Six shillings for the Spanish dollar became thenceforth the standard of lawful money in the colonies, although the currencies of some of them afterwards departed from it. In 1717 Sir Isaac Newton, being master of the mint, again made assays of the Spanish dollars, and found them still to contain 387 grains. From this standard they successively fell off in 1731, in 1761, and in 1772; since which their average weight and purity has been that at which the dollar of the United States is fixed.

The dollar being thus of the intrinsic value of four shillings and six pence sterling, the pound sterling was of course equivalent to 4 and $\frac{4}{9}$ of the dollar. This was the par of exchange, computed in the *silver coins* of the two countries, for even then if the computation had been made between their gold coins, the result would have been different.

Thus, while the laws of the United States, in establishing their mint, and the unit of their currency, have assumed for their standard the Spanish dollar of 1772, in the calculations of their revenue and their estimate of the English pound sterling, they have adopted the Spanish dollar of 1704.

But when, in 1704, the value of the Mexican dollar was fixed at four shillings and six pence, it was because it contained 387 grains of pure silver, the same quantity which was also contained in four shillings and six pence of English coined silver. At this time, four shillings and six pence sterling of English silver coin, contain only 363¼ grains of pure silver, and the dollar of the United States contains 371¼ grains.

The following statements show the relative present value of the dollar and pound sterling in the gold and silver coins of both countries, in gold bullion, as payable by the Bank of England, and in English bank paper at its current value in 1815.

1. *Gold.*

One pound troy weight of standard gold in England contains 5,280 grains of pure gold. It is coined into £46 14s. 6d. or 11,214 pence.

Then 11,214 : 5,280 :: 240 : 113.0014 grains of pure gold in a pound sterling.

In the United States 24.75 grains of pure gold is coined into a dollar, or 247.5 grains to an eagle.

Then 24.75 : 1 :: 113.0014 : 4.56572 dollars, cents, &c. to £1.

Thus the pound sterling in gold is worth $4 56.572.

And as 5,280 : 11,214 :: 24.75 : 52.5656.

Dollar in English gold 4s. 4.5656.

Pound sterling in gold $4 56.572.

2. *Silver.*

One pound troy weight of standard silver in England contains 5,328 grains of pure silver, and is coined into 66 shillings, or 792 pence.

The dollar of the United States contains 371.25 grains of pure silver.

Then, 5,328 : 792 :: 371.25 : 55.1858.

Dollar in English silver 4s. 7.1858.

792 : 5328 :: 240 : 1,614.545 grains pure silver in a pound.

371.25 : 1,614,545 :: 1 : 4.348943.

Pound sterling in silver $4 34.8943.

Medium par dollar, 4s. 5.8757 pence.

£ stg. in gold $4 56.5720 —
    in silver 4 34.8943 +   10.8388 = $4 45.7331 med. par £ stg.

3.

Value of the pound sterling and dollar in gold and silver coins, in gold bullion, and in English bank paper.

	Pence stg.
Value of United States dollar in English silver coin at 66 shillings per lb. troy weight -	55.1858
In English gold coin at £3 17s. 10½d. per oz.	52.5656
In English bank notes in 1815, - - -	72.
In gold bullion at £4 1s. per ounce, -	54.675

	D. Cents.
English pound sterling, in silver coin, worth in the United States, silver dollars, -	4 34.8943
Gold coin at £3 17s. 10½d. per oz. in United States gold,	4 56.5720
In English bank notes, 1815, - - -	3 33.3333
In gold bullion at £4 1s. per ounce,	4 38.9574
In ditto at 4, - - -	4 44.4444

APPENDIX.

D 1.

Mr. Hassler to the Secretary of State,

NEWARK, N. J. 16*th* October, 1819.

MOST HONORED SIR: When I had the honor to see you relative to my appointment for the boundary line, the conversation falling upon French and English standards of weights and measures, I mentioned to you that I had brought a complete set of them with the instruments for the survey of the coast, of which I handed you a catalogue.

Mentioning at the same time that I had made comparisons of those standards of length-measures, and some others which I had besides, (having not had time to extend it to the standards of weights.)

Since then, I could not occupy myself with this subject, until lately in the course of the papers upon the scientific part of the survey of the coast, which I prepared for publication.

I have, therefore, now the pleasure to fulfil my promise, and your desire of that time, by forwarding you herewith an extract of this paper, containing the part which may interest you, with the necessary details to convey the conviction of accuracy necessary in such subjects to inspire confidence in the results.

You will observe that these are not so full and extensive as I intended them to be. I am sorry that the circumstances which befel this, in a national point of view, so honorable and useful work, have also in this part frustrated the aim of my exertions.

I believe however, what has been obtained will be sufficient for your purpose of a report to Congress upon this subject.

If you should wish any thing more that I could be able to do, I will do it with pleasure.

You may probably conclude with me from the comparisons of Captain Kater and other circumstances, that a platinum *meter copy* will yet remain much inferior in real scientific value, to the iron original of the committee which I had in trial, and that the large expansion reduction necessary in the comparison by the English way of giving the ratio of the standard, may introduce some uncertainty by the less accurately known expansion of platinum. I have, besides, repeatedly, and, also, in these comparisons, had some reasons to suspect that copies are most generally shorter than the original in standards cut to a determined length.

    I have the honor to be,
      With perfect respect and esteem,
       Most honored, Sir,
        Your most obedient humble servant,
         F. R. HASSLER.

The honorable JOHN QUINCY ADAMS,
  Secretary of State to the United States,
    Washington City.

D 2.

Comparisons of French and English standard length measures made in March, 1817, by F. R. Hassler.

The two length measures which have been the most scientifically ascertained and compared, are the French and the English.

They are essentially different in their principles, and of different metals, which circumstance has always presented difficulties in their comparison.

The English standard is *a brass scale of undetermined length,* divided into inches and tenths of inches. They are of different ages and accuracy, and successively perfectioned by various artists, by making new scales of any convenient length from the mean of distances taken upon the old scales. Sir George Shuckburgh Evelyn's account of comparisons of a number of them may be consulted for the nearer details, (Philosoph. Trans. of London.)

The French standard is *a certain determined unit of length in iron,* given by a bar cut off to the given length, either a toise, or a meter. The accounts of their comparison, ratio, and the determination of the latter from nature, is contained in detail in the Base du Systeme Metrique.

Whoever has attempted to copy an absolute length, or to multiply it with accuracy, to form a long standard from a short one, will soon have discovered that great care is needed in the choice of means. Beam-compasses and similar means are fully unsatisfactory for both. This conviction, and the perusal of the different modes of proceeding used in Europe in the works of this nature, decided me to adopt for the unit measure of the bars, to be used in the base-measuring apparatus, the mechanical combination of four iron bars, of two metres length each, (of which account is given in another part of the papers,) an additional deciding reason for this was, that I had to my disposition for this use one of the metres standarded by the committee of weights and measures in Paris in 1799, which is therefore equally authentic with the one in Paris itself, and places the accuracy of my unit measure above all possible doubt, while of any other measure I could only obtain the copy of a copy. The comparison of this with the other copies mentioned in this paper, and my bars, rendering it besides comparable to any standard by mere numerical calculation.

The comparisons to be related here were made in February and March, 1817, and intended to be repeated before the measurement of the first base line of the survey of the coast; which, having become impossible, by events which I was too far from supposing possible, I will here only give the result of what was done.

The following are the different standards compared and their origin, which is the first thing to be related minutely :

1st. An iron metre standarded at Paris in 1799, by the committee of weights and measures, composed of members of the institute and foreign deputies, ad hoc,—its breadth is 1.″1 Engl. its thickness =

0″37, like all metres then made, to which it is exactly similar, it bears the stamp of the committee, namely, a section of the elliptic earth, of which one quadrant is clear, with the number 10,000,000 inside of the arc, the other three quadrants being shaded.

My friend, Mr. J. G. Tralles, now member of the academy of Berlin, was at that time deputy of the Helvetic republic for this purpose, and, as appears by the account of the proceedings, the foreign member who attended the construction and comparison of the length measures. He was so kind as to have this metre made expressly for me simultaneously with all the others, passing in all respects the same process and comparisons; and on his return to make me a present of this, as well as of a kilogramme, also standarded by the committee, under the direction of Mr. Van Swinden, (being No. 2.)

2d. A very well executed iron toise, with its mother, in which it fits, forming with it a bar of three inches broad and half an inch thick. It is made by *Canivet a la Sphere a Paris;* this is engraved upon it, together with the notice, " *Toise de France étalonée le* 16*me* 8*bre,* 1768, " *a la temperature de* 16° *du thermometre de Mr. la Réaumur.*" On the back edge of the toise a line is drawn in the middle over the whole length, and, from a perpendicular crossing this line near one extremity, a point is laid off near the other extremity; along this line is engraved " *La double longueur du pendule sous l' equateur,*" a point being also in the middle, at the simple length of the pendulum.

Having been in Paris in 1796, the heirs of Mr. Dionis du Séjour, who had died shortly before, had given this toise to the celebrated artist, Mr. Lenoir, to sell it, from whom I bought it, considering it as the best and most authentic standard of this kind in private hands.

It is well known that about the time stated for the standarding of this toise, the Academy of Sciences had in contemplation to establish as a natural standard the double length of the pendulum under the equator, marked on this toise, probably to this very purpose, in which Mr. Dionis du Séjour must, by his situation at the academy, have taken particular interest. The work denotes its intention to a valuable purpose.

3d. Two copies of the toises of Mr. Lalande, which have been compared in England with Mr. Bird's scale of equal parts in 1768, after the return of Messrs. Mason and Dixon from the measurement of the degree in Maryland. These copies being of the same size and shape as the originals.

When I was in Paris, in 1793, Mr. Lalande communicated to me the above toises, for the use of the survey of Swisserland, which I had then begun. Mr. Tralles and I made two copies of each of them, those here compared being one set. The toises of Mr. Lalande are known to be marked A and B, but only on the woods in which they are framed, and from which they can easily be changed: our's were marked upon the iron itself, so as we found them, in the time, on the toises of Mr. Lalande.

N. B. The standards hitherto mentioned, I brought with me to this country at my first arrival in 1805, and ceded them some time after to John Vaughan, Esq. of Philadelphia, who has been so good as to lend them to me for the intended comparison.

4th. The brass metre standarded by Lenoir, No. 16 of the collection of instruments for the survey of the coast, which was compared at the Observatory of Paris, as per certificate of Mr. Bouvard and Arrago, its breadth is $=1.''1$ English, its thickness $=0.''18$. The certificate, dated at the Observatory, 16th March, 1813, says of it: " En applicant a nos mesures une correction dependante de l'inega- " lité de dilatation des deux metaux, il nous a semblé qu' a zero du " thermometre (centigrade) le metre en cuivre de Mr. Hassler, se- " roit plus court que l'Etalon en fer de nos archives de $\frac{1}{100}$me. de " millimetre."

Mr. Lenoir, who has been employed by the comitée of weights and measures to construct the standards, had taken the precaution, at that time, to make one brass metre, which passed all the comparisons with the others, at the standard temperature of 0° centigrade, at which, therefore, it is equal to the authentic iron metres, and forms the only direct mean of comparison between French and English measure. Of this metre the present is a copy, which I thought so much more interesting to have, as the original is unique in its kind, and the comparison of it with the English standard possible by direct means.

5th. One iron toise of Lenoir, No. 15 of the collection of instruments for the survey of the coast. It is near two inches broad, and about a fourth of an inch thick. Its comparison at the observatory at Paris, by the certificate quoted above, says it to be exactly equal to the toise of Peru, preserved at the said observatory.

6th. One iron metre, standarded by Lenoir, No. 18 of the catalogue of instruments for the survey of the coast. It is of the same breadth and thickness as the iron metres of the comitèe, but was not compared at the observatory of Paris, on account of being received too late.

7th. An iron bar, similar to the metre just mentioned, which I intended to bring to the proper metre length, for myself, being yet too long. It was used in the comparison; as well as an operation necessary for its standarding, as to vary the means of combination in the comparisons. It has not been worked since this, having been reserved for a future comparison.

8th. A brass standard scale, of English measure, of 82 inches, divided on silver to every tenth of inches, made by Mr. Troughton, No. 14 of the catalogue of instruments for the survey of the coast. It has the arms of the United States engraved upon it, and " *Troughton, London,* 1813." It is three inches broad, and half an inch thick.

To it belongs an apparatus for comparing measures by two compound microscopes, sliding on a rule of equal breadth and thickness with the standard ; and adjustable parallel to it one microscope, having a micrometer reading directly the $\frac{1}{10000}$ of an inch.

APPENDIX. 157

A description of this arrangement existing already in Nicholson's Journal, though on a much smaller scale, all details may here be omitted.

The scale was divided by Mr. Troughton with the utmost care and accuracy, so justly praised in this eminent artist. It contains the double length of the principal part of his own scale, of which an account has been given in the paper of Sir George Shuckburgh, above quoted. Mr. Troughton compared his scale first again in itself, and made a table of errors for the same, in like manner as he has described in his method of hand dividing (Philosophical Transactions, 1809) and then laid off the scale here used, correcting each point according to this table.

To give to a bar of a certain thickness an accurate length, by a cut perpendicular to its length, is an operation which cannot be performed by the free hand, which will always work it uneven, and most likely rounded. To do this with accuracy I had a tool constructed, &c. &c.

(The description of this tool, the iron bars, and the manner in which they have been worked, may here be passed over, merely mentioning that the four bars were of equal breadth and thickness with the metres, and lettered A, B, C, D.)

For the actual comparison, the unequal thickness of the different standards obliged to support them, so as to bring them exactly to the elevation of the thickest, for which the foci of the microscopes are adjusted, and the value of the micrometer determined. In the comparison of the metres the brass scale was the thickest, and in that of the toises the toise of Canivet. The influence of this necessary supporting, when done only partially, is very great, therefore I had pine rules, made of sufficient breadth and length to support fully the standards and the butting pieces, which were used in reading off, as will be said immediately, and of the accurate thickness to bring each of the metres exactly to the focus, and free from all parallax. As it is completely inadmissible to take the edge of a bar, if ever so sharp, as object under the microscope for the purpose of comparison, because it never forms a good image, pieces cut off from the end of the above mentioned bars, (which had originally been made purposely about nine inches too long) were placed at the two ends of the metres, under comparison, to make the reading from the separation lines, presented by their close contact, which were not thicker than the lines of the scale. These pieces were from two to four inches long, worked to the exact thickness of the metre, or toise, with which they had to serve, and by rubbing with emery and oil one against the other, turning them in all positions, they were brought to make exact contact on the whole surface, which afterwards joined to the metre did the same with it, and on account of their equal thickness presented a plane crossed by the partition line. Upon this the micrometer wires which cross each other under an angle of about 30°, were brought, by optical contact, so that the partition line bisected the angle exactly in the crossing point of the wires.

APPENDIX.

The microscopes were furnished with paper reflectors, by a piece of white paper placed in a forward inclined direction between the microscopes and their supports, by which the light was reflected upon the scale or standard in the direction of the division lines, as required to avoid all shades of the lines, or partitions, which prevent accurate reading.

To prevent the influence of the heat of the observer's body upon the scale and apparatus, a large sheet of paper was nailed to the work bench, near the microscopes, where the observer approaches, and I worked with gloves on for the same reason.

From seven to twelve thermometers were constantly laying over the scale and the standards, and read off at proper intervals of time.

The workbench itself was near double as long as the scale, to obtain a sufficient length fully accurate plane, it was in this respect accurately adjusted before the work, and so placed in respect to the windows, that each microscope corresponded to the best equal light from a separate window; the bench was made of two three inch planks at right angle, and upon six legs.

There was no fire made in the room during the comparison; and, for a number of days before, the windows were besides left open day and night, to bring the room to a steady temperature, being equal to that of the atmosphere.

For the intended comparison of a day all was prepared the day before, and left lying just fit to begin the observations the next day, in order that all parts might be fully at rest, and after verification in the morning the actual comparison begun.

All these precautions are necessary to obtain satisfactory results, as I think may be observed by what will follow.

The possible inaccuracy of the readings may be considered in the direct ratio of their number, having four metres, and the scale holding the sum of two, I had the mean to halve this error by comparing always two together alternately, reading their sum at once on the scale, and forming an equation between the results, to obtain the value of each individual metre. This method had yet the advantage of leaving the observer fully unprejudiced upon what he shall read, as the combination of the different measures and the different influence of temperature occasions a variation which precludes previous estimates.

To this effect, the microscopes were placed to the decimal on the scale nearest to the sum of two metres, viz: $78.''7$ or $78.''8$, which was taken from $+1''$ to $79,''8$, which approximated it nearest to the middle of the scale. When the scale had been removed, and the two metres with their support properly laid under, so as to bring the middle of their breadth under the centre of the microscopes, the middle contact was exactly made, and with a magnifying glass verified, then the butting pieces being laid on, the coincidence of one end with the crossing wires of the fixed microscope was obtained by longitudinal motion, and the micrometer wires being moved upon the other line of contact, between the metre end and the butting piece, the va-

APPENDIX. 159

lue of the corresponding subdivisions was read on the micrometer, by its revolutions and subdivisions. The longitudinal motion cannot be made by hand; it is obtained by light strokes with a proper piece of wood, and requires dexterity and care, particularly not to separate the different pieces by the counter stroke.

The value of the micrometer parts was to be previously ascertained by accurate and repeated measurements of a decimal on the scale in different places; by a mean of many, I found under the adjustment for the metres $0.''1$ on the scale $= 1.''004$ of the micrometer.

At last the individual value of the distance used on the scale in relation to the mean value of the same distance resulting from the measurement of it, on as many parts of the scale as was admissible, was to be determined, in order to give the true mean value of the distance of the points of the scale compared. This was done by about 50 measurements principally after the comparison had been made, and before any alteration had been made in the microscopes. So I found the distance used, or $79.''8 - 1.''0 = 78.''800172$ of the mean value of the scale; to this, therefore, all the values obtained by the metre comparison were ultimately referred.

To shorten the manner of registering the results and the combination of the metres and their position, the following mode of notation was adopted:

Mc denotes the iron metre of the committee of weights and measures.
Ml Lenoir.
Mb brass metre of Lenoir.
Mj iron bar which I intended to bring to a metre length.
M$c+l$ -   metre of the committee and that of Lenoir together, all marks up.
M$ɔ+\jmath$ -   the same metres, all marks down.

And, in like manner, in the other combinations, the adding of the special marks at the top, denoting always the sum of the metres so indicated, and the inversion of the letters the inversion of these metres.

The 15th March, early in the morning, the eleven thermometers which had been lying over night on the scales, prepared the evening before, were read; after having verified that all was in order, then it was further observed, as by the following table:

160 APPENDIX.

Standards under comparison.	Measures of the micrometer.	Mean of the four results.	Correction for the micrometer value.	Final value of the mean.	Mean of thermometer, Fahrenheit's temperature.
Mc+b	78.″76040				
Mɔ+q	.76115				
Changed ends for middle.					
Mc+b	78.76099				
Mɔ+q	.76131	78.760962	—0.″000244	78.76078	30.°85
Mb+l	78.75903				
Mq+ɿ	—.760575				
Changed ends for middle.					
Mb+l	78.75920				
Mq+ɿ	—.760303	78.759777	—0.000240	78.759537	
Mc+l	78.760415				
Mɔ+ɿ	—.760450				
Changed ends for middle.					
Mc+l	78.760475				
Mɔ+ɿ	—.76055	78.760472	—0.000242	78.60230	34.1

The four bars intended for the base measurement were now also compared, bu the result is omitted as unimportant here.)

The micrometer microscopes having in the foregoing comparison, read by addition, or from 78.″7 onwards, i was, the 7th March, turned for one half revolution horizontally, so as to read by subtraction, or from 78.″8 backwards, in order more effectually to compensate the influences possible from them by the following comparisons. The micrometer values were verified, and found as before. Then all was prepared for the comparison of next day. All other things being otherwise equally disposed.

APPENDIX. 161

The 8th March, A.M. all having been verified, the comparisons were made as by the following Table:

Standards compared.	Microscope Readings.	Mean of the Results.	Correction for Microm. ft.	Corrected Rdl of Reading.	Final ft.	Mn Temperatr.
	78."8					46.6
M =	0.044075				+	
M₃+₁=	0.043000					
Changed ends for middle,		0.043900	0.000 756	0.043724	78.756276	
Mc+l =	0.043250					
M₃+₁=	0.043235					
Mc+j =	0175					
M+l =	0.039300					
Changed ends for middle,		082	000. 588	0.039643	—.760357	
Mc+j =	0.040200					
M₃+l =	0.040135					
Ml+j =	0.041525					
M₁+l =	0.040830					
Changed ends for middle,		0.04 295	0.000165	0.041130	—.758870	8.6
Ml+j =	0100					
M₁+l =	0.041225					

The four double Metal Bars were now successively put under comparison in their four possible positions and found —

Changed end for end,		0.044369	0001775	0.044 9 5	78.7568085	
A =	0.044475					
V =	0.044250					
A =	0. 0425					
V =	0. 0425					

162 APPENDIX.

Standards compared.		Mn of the four Results.	Corrections for M-m. Value.	Corrected Result of &c.	Final Value.	Mean tm. Ha-
B =	0.0437 5					
B =	0.043750					
Changed end for end, -		0.043362	0.0001734	0.043 886	—.7568114	
B =	0.043150					
B =	0.043175					
C =	0.043575					
C =	0.0475					
Changed end for end, -		0.043794	0.0001752	0.0436.9	—.75638	49.8
C =	0.0350					
C =	0.043975					
D =	0.043300					
D =	0.043100					
Changed end for end, -		0.0433625	0.000 7345	0.043 890	78. 56811	50.5
D =	0.043550					
D =	0.043500					

The same day P. M. the comparisons were made as follow:

M+l =	0.0490					
M+l =	0.0425					
Changed ends for middle,		0.044829	0.0001793	0.0446497	78. 36	50.3
M+l =	0.0450					
M₂+l =	0.0450					

APPENDIX. 163

$Mc+j=$	0.0475		
$Mo+l=$	0.040525		
Changed ends for middle,	0.040850	0.040850	
$Mc+j=$	0.040850		
$Mo+l=$	0.041550		
$Ml+j=$	0.043175		
$M_2+l=$	0.042450		
Changed ends for middle,	0.042300	0.0426125	
$Ml+j=$	0.042300		
$M_2+l=$	0.042525		

To make these results … able, it is n… ay to de… tim al to one t… ture by the difference of expansion between iron and bs. I will for this make use of my … ults of the pyrometrical experiments made immedi… ly after the … on, and published in the philo… ical transactions of Philad lph a of the same year, he esults of which … ge he expansion for one degree of t r mp ra ure of … Fahrenheit's sca e, 0.0406866 —.7593 3

Difference.
$$\text{In iron} = 0.00000696535 $$
$$\text{In brass} = 0.00001050903 \bigg\} 0.00000354495$$ Dec ma pa ts of the eng h.   0.042442   —.7575579

On a o unt of the ch nge of temperature during the … rim, and the … the influence of this elemen in the results, it is necessary to ake for ach comparison the proportional temperature corresponding to it ; bu n the readings of the thermometer the work … pared ng r gu arly, ea le single compar son will … espond to the … ortional t m…

The standard mpr ture to hich I find it most … tural to reduce h e measurements is 32° Fahr. or 0° c ntesimal & Reaumur, it being … ind f r the me er and … te. I will reduce the results of the iron to … ws, so th t the numb rs will express the length of the … ters in English nches of the b ass scale, both at the temperature of 32°, which appears to me most naturally, being possible to obtain by actual experiment, while the giving of the meter at 32° in ength, on the scale at 62°, is impossible to produce and … er fy in nature, … ere … w ys a … s lt of … la on in which the atio of expans on used las too … ch 1 … . The b ass m re s ther c e ns … ed as needing no … n.

51.8

The constant quantity of 0"000 72 is also to be added to each … me … ch n on … cant of the ind v du l va ue of he part of the scale made use of.

APPENDIX.

The following table will therefore present the results of all the foregoing comparisons, with their reductions:

Date of comparison.	Standards compared.	Temperature of the comparison.	Result of the comparison.	T. 32°	Reduction for temperature.	Value at 32° F.
5 h A.M.	$Mc+b$	3°.4	78.7607 8	0°.6	0.00083773	78.760806226
	$Mb+l$	32.5	—.759537	+0.5	+0.000069811	—.759778811
	$Mc+l$	33.7	—.760230	+1.7	+0.000747134	—.760876713
18th March, A. M.	$Mc+l$	47.1	78.756276	15.1	0.0042165	78.7606645
	$Mc+j$	48.1	—.760357	16.1	0.0044957	—.7650247
	$Ml+j$	48.9	—.758870	16.9	0.0047191	—.7637611
	Bars A	49.2	78.7558085	17.2	0.00480286	78.76078336
	B	49.5	—.7568114	17.5	0.00488162	—.76186502
	C	49.8	—.7563810	17.8	0.00497040	—.76152340
	D	50.2	—.7568110	18.2	0.00508204	—.76206509
8 h P. M.	$Mc+l$	50.5	78.7553503	18.5	0.005 6567	78.76068797
	$Mc+j$	51.0	—.7563 54	19.0	0.0053055 5	—.7647909 5
	$Ml+j$	51.5	—.7575579	19.5	0.0060 10	—.7631750

By the principles of the arrangement, it is evident that the value of any one single one compared obtained by a simple equation of the form

$$C = \frac{(c+b) + (c+l) - (b+l)}{2};$$

and so any of the others, mutating the letters accordingly.

APPENDIX.

The final results of these comparisons form therefore the following table of the values of the different meters compared at the temperature of 32° of both metre and scales, in English inches:

Date of the comparison.	M*c*.	M*l*.	M*b*.	M*j*.
15th March,	39.380952064	39.37992415	39.379854162	
18th March, A. M.	—.3809641	—.3797004		39.3840606
P. M.	—.38115196	—.37953601		—.3836290
Means	39.381022708	39.37972015		39.3838448
Correction of the brass metre by the certificate	-	-	+0.00039381	
which applied, gives the metre corrected			39.38024797	

These results might now be compared with those obtained by various comparisons made in England, these being however always stated so as taking the metre at 32°, and in value of the English scale at 62°, it is necessary to reduce them all for 30° difference of temperature full expansion of the brass. As I have not now the books in which they are related, and am ignorant, so various are they, which English standard and expansion has been used; supposing, however, that it has most generally been that of Borda, I will here only present, in a tabular shape, the different results as I have them, and reduce them to 32°, to compare them with my results. Observing, at the same time, that they are yet subject to the differences between the English standards themselves, which are in some instances greater than the differences of these results, as may be seen by the paper of Sir G. Shuckburgh, quoted above, and the account of Mr. Pictet, of Geneva, made in London in 1802. Borda's expansion for brass being 0.00000999, (though I have seen it lately stated at 0.0000101, on what ground I do not know, unless I suppose a mistake.)

Authority or observer.	Value given at 62°	Reduced to 32°.	Difference with my results of committee metre
The Roy. Soc. accepted in 1800	39.370572	39.38126801	+0.0002453
	—.3702	—.380896	—0.0001267
Mr. Pictet in 1802	—.371	—.381696	+0.0006733
Mr. Kater on S. G. Shuckburgh's scale, lately, 1818	—.37079	—.381486	+0.0004633
The same on Bird's scale (Unknown which of Bird's scales, there being a difference.)	—.37062	—.381316	+0.0002933

I do not compare by my ratio of expansion, because they were not made or known at the times of the older comparison; of course could never have been employed in them.

The 21st March, I took the different standards of the toise under comparison.

The toise of Canivet being half an inch French in thickness, and the brass scale half an inch English, this difference was compensated by laying four thicknesses of white paper strips under the whole length of the scale; the microscopes were adjusted to fit this toise, and then the scale adjusted to it; the other toises had rules of proper thickness to bring them to the same focus.

The distance of 76."8, as nearest to the toise was taken between the microscopes, from + 2" to 78."8 on the scale as bringing the measure again nearest the middle of the scale.

The value of the micrometer was determined by repeated measurement of the decimal on the scale between 78.7 and 78.8 and found that it measured 0."10053 by the micrometer; the readings here given, will, therefore, be corrected by this value of the micrometer, by which they will represent regular decimals of this individual subdivision.

It was of course also here intended to compare the distance taken on the scale with all the other measures of the same distance which can be measured on the scale, and the value of the micrometer upon a great number of decimal divisions, to obtain a mean value of it. But it being necessary to begin the pyrometrical experiments before the cold weather should cease, I delayed this comparison until after, or to another opportunity, which did not occur before I delivered the whole of these standards, and the scale, with the other instruments. The values here obtained will, therefore, remain individual, unless some future occasion should present to make this necessary trial of the scale on the indicated length, when the results here given may be corrected accordingly.

The microscopes being screwed fast, and the 0.° of the micrometer not agreeing fully with the division of the scale, which, for small differences it is not proper to correct by the screw of the sliding piece, I used the better method of reading repeatedly the divisions of 78.7 and 78.8 on the micrometer, taking the point of the micrometer so determined, by a mean, for the zero, from which the readings of the micrometer are to be subtracted, as the micrometer read yet by subtraction from 78."8, this point being 78."8001375, all the readings given immediately, are to be subtracted from the number instead of 78."8 only.

I intended of course also to repeat the comparison with the microscope reading directly or onwards, but the same reasons stated prevented the execution of it.

Using again abridged notations for the registering of the results, I called C = the toise of Canivet
    L =   Lenoir
    l^A =   Lalande marked A
    l^B =   -   -   -   B
} inverting the letter when the marks laid downwards.

In the toise of Canivet, three points were observed on the length of the contact, lettered as follows:

i, at ¼ of the contact from the inner corner
m, in the middle of the contact
e, at 0."05 from the outer end of the contact

In the toise of Lenoir only the middle was taken, (which is marked by a line,) like in the comparison at Paris. The same was done with the toises of Lalande. The turning end for end of the toise of Canivet in both cases related was impossible, on account of its breadth. These comparisons present the following table:

APPENDIX. 167

Toise compared.	Temp. can.	Meter readings.	Means of the readings.	Corrections for microscopic er.	Corrected dng.	Final v ud.
	33.°4	6,800.375		—0.00053	—	
C_m		0625				
C_i		5050				
C_e		0640				
		0.05705 0.058675 } 0.0573417 } 0.0576958		0.0003058	0.0573900	76.7427475
C_e						
C_i						
C_u						
	35.°	053				
L		0.06055				
T		changed end for end .6062		0.0003 3	0.0592987	76.7408389
L		0.06070				
T		00				
	36.°1					
A		06102 } 0.061835		0.0003277	0.0615073	76.7386302
A'		0.06265				
B		0.0536 } 0.053675		0.0002845	0.0503905	76.7 67470
B'		0.05375				
	39.°0					

The results of the foregoing table are now to be reduced to the standard temperature of 32° by the difference of expansion of iron and brass, for which I shall again use the result of my own experiments. They present by the following table of results of toise comparisons:

Denom. of Toise.	Temperature.	Result of the comparison unreduced.	T. 32°	Reduction for temperature.	Value at 32°.
C	34.2	6.″742 4 5	+ 2.°2	0.00059722	76.″74334472
L	36.0	−.7408389	+ 4.0	0.00108820	−.74192710
lA	37.4	−.7386302	+ 5.4	0.00146921	−.74009941
lB	38.5	−.7467470	+ 6.4	0.00174145	−.74848845
$\frac{l\text{A} + l\text{B}}{2}$					

as accepted in the comparison of the original of the toise in .768, in London, 76.74429393

APPENDIX.

Having no books about me in which I could see for the details of comparison of toises with English measures, the comparison just quoted is the only one with which I could compare mine, and the inferiority of the two toises of Mr. Lalande in every respect renders it of very little importance.

In the comparison of 1768, the mean adopted as result was $\frac{lA + lB}{2} = 76.734$ at the temperature of 62°, which is stated to be 0.024 of an inch longer than when determined by Mr. Graham. I have not before me sufficient data to judge of the propriety of the mean adopted, as it does not follow immediately from the two individual results stated in my notice. In respect to the temperature it seems that both toises and the scale of Bird of equal parts were supposed both at the temperature of 62°. Reducing this result to 32°, by different ratios of expansions, it presents the following result comparable with mine:

	Ratio of difference of Expansion.	Amount of Expansion, +	Value at 32°
By Borda's expansion	0.00000352	0.0081031	76.7421031
By Troughton's	0.0000039	0.0089779	—.7429779
By my experiments	0.000003545495	0.0081618	—.742162

Which are all smaller than mine, and I should suppose to deviate more from truth, as they would give a more erroneous proportion between the metre and toise. But I must again observe, that my results may yet be subject to a small correction by the reduction of the distance used on the scale to its mean value, as has been done with the metres.

The ratio between metre and toise, which may be quoted, as well for curiosity' sake, as also as a kind of trial of my comparison, may here be stated in all combinations which the different standards admit; and I have in doing so the private interest of giving the ratio of the toise of Canivet to an authentic metre; because this toise has served in 1796, to Mr. Tralles and myself, to measure a base line of more than 40,000 feet by an arrangement of toises bars, which were standarded in it.

The ratio of the metre to the toise, or the metre expressed in decimals of the toise, is given by Mr. Delambre.

Metre = 0.513111185

With my denomination is obtained:

Toise

$$\frac{Mc}{C} = 0.513152$$

$$\frac{Mc}{L} = 0.513162$$

$$\frac{Ml}{C} = 0.5131355$$

$$\frac{Ml}{L} = 0.513145$$

$$\frac{Mb}{C} = 0.513137$$

$$\frac{Mb}{L} = 0.513146$$

$$\frac{Mc}{l\frac{A+B}{2}} = 0.513146$$

$$\frac{Ml}{l\frac{A+B}{2}} = 0.513130$$

$$\frac{Mb}{l\frac{A+B}{2}} = 0.513131$$

Captain Kater's comparison of the metre with Bird's scale would give

$$\frac{M}{C} \overset{\text{B. Sc.}}{=} 0.513156$$

$$\frac{M}{L} \overset{\text{B. Sc.}}{=} 0.5131658$$

With sir George Shuckburgh's scale the deviation would be still greater.

F. R. HASSLER.

Newark, New Jersey, October, 1819.

E 1—*a.*

NEW HAMPSHIRE.

Governor of New Hampshire to the Secretary of State.

EXECUTIVE DEPARTMENT,
Epping, November 21, 1818.

SIR: By the last mail I received your letter of the 4th instant requesting a copy of the law of New Hampshire of 1718, upon the subject of weights and measures, and a statement of each scale beam, weight, and measure, described in the act of 1797, and where, and in whose custody, they are kept.

I herewith send you a copy of the law of 1718, and of an additional act passed the 6th of George III. upon that subject. I transcribed them from books in my library; and, as I presume you do not wish to use them in a court of law, I did not suppose it necessary to send to the Secretary of this State for official copies, especially as I doubt whether even copies remain of both those laws in that office.

I have taken measures to obtain the requisite information respecting the scale beams, weights, and measures, belonging to this state, and as soon as I obtain it will write you on the subject.

I have the honor to be,
With much respect, sir,
Your obedient servant,
WILLIAM PLUMER.

Hon. JOHN QUINCY ADAMS,
Secretary of State U. S. Washington.

E 1—b.

Copy of an Act passed by the General Court or Assembly of the Province of New Hampshire, in New England, begun and held at Portsmouth, on the 13th day of May, 1718.

AN ACT FOR REGULATING WEIGHTS AND MEASURES.

To the end that weights and measures may be one and the same throughout this Province,

1. *Be it enacted by his Excellency the Governor, Council, and Representatives, convened in General Assembly, and by the authority of the same,* That the treasurer of this province shall provide one set of weights and measures as are according to the approved Winchester measures, allowed in England in the exchequer, which shall be public allowed standards throughout this province, for the proving and sealing of all weights and measures thereby: and the town clerk of every town within this province, not already supplied, shall, within three months next coming, provide, upon the town charge, one bushel, one half bushel, one peck, one ale quart, one wine quart, one ell, one yard, one set of brass weights to four pounds after sixteen ounces to the pound, with fit scales and steel beam, tried and proved by the aforesaid standard, and sealed by the treasurer, or his deputy, in his presence; which shall be kept and used only for standards in the several towns, who is hereby authorized to do the same, for which he shall receive, from the town clerks of every town, two pence for every weight and measure so tried, proved, and sealed. And the town clerk of every town shall commit the weights and measures unto the

custody of the select-men for their town, for the time being, who, with the town clerk, are enjoined to choose an able man for a sealer of all weights and measures for their own town, from time to time, and until another be chosen, who shall be presented to the next court, or some justice of the peace, to be sworn to the faithful discharge of his duty, and shall have power to send forth his warrants by the constables to all the inhabitants of each town, to bring in all such weights and measures as they make any use of, in the month of April, from year to year, at such time and place as he shall appoint, and make return to the sealer in writing of all persons so summoned; that then and there all such weights and measures may be proved and sealed, with the town seal; which is likewise to be provided by the town clerk, at each town charge, who shall have for every weight and measure, so sealed, one penny, from the owner thereof at the first sealing. And all such weights and measures as cannot be brought to their just standard, he shall deface and destroy: and, after the first sealing, he shall have nothing so long as they continue just with the standard.

2. *And it is further enacted by the authority aforesaid*, That, if any constable, select-men, or sealers, do not execute these laws, so far as to each and every of them appertains, shall forfeit for every neglect the space of one month, the sum of forty shillings; the half to the informer, the other half to the use of the poor of the town where such default is found; and every person neglecting to bring in their weights and measures at the time and place appointed, being only warned thereof, shall forfeit three shillings and four-pence; the one-half thereof to the use of the poor, as aforesaid, the other half to the sealer; and the penalties herein mentioned to be levied by distress, by warrant from any justice of the peace within this province.

3. *And it is further enacted by the authority aforesaid*, That, in every seaport town within this province, the town clerk is to provide, likewise, upon the town charge, one hundred weight, made of iron, to be tried, proved, and sealed, as aforesaid, and one half hundred, and one quarter of an hundred, and one fourteen pound weight, made of iron, to be tried, proved, and sealed, as aforesaid, and to be kept as standards in the said several towns, to be used as before for other weights and measures as is directed.

4. *And it is further enacted by the authority aforesaid*, That all steelyards that are or shall be approved of by the standards, shall be allowed of in any of the towns of the said province, and be in the liberty of both buyer and seller to weigh by which they please.

5. *And be it further enacted by the authority aforesaid*, That all measures by which meal, fruits, and other things, usually sold by heap, shall be sold, be conformable, as to bigness, to the following dimensions, viz: The bushel not less within side than eighteen inches and half wide, the half bushel not less than thirteen inches and three quarters wide, the peck not less than ten inches and three quarters wide, and the half peck not less than nine inches wide. And if any person, at any time, from and after the first day of October next,

after the publication of this act, shall sell, expose to sale, or offer any meal, fruits, or other things usually sold by heap, by any other measure than is aforementioned, as to bigness and breadth, such person being complained of and convicted before any justice of peace within this province, of so doing, shall forfeit and pay, to the use of the poor of the town where the offence is committed, the full value of the meal, fruits, or other things so sold, or offered to sale; and such justice may commit the offender to prison, until payment be made of the said forfeiture, or cause the same to be levied by warrant of distress, and paid unto the town treasurer, or overseers of the poor, to the use of the poor aforesaid; and shall also cause such measure to be defaced: any law, usage, or custom, to the contrary in any wise notwithstanding.

6. *And be it further enacted by the authority aforesaid,* That the sealer appointed in each town within this province, from time to time, shall be, and hereby is, empowered to go to the houses of such of the inhabitants as, upon warning given in manner as is above appointed, shall neglect to bring or send in their beams, weights, and measures, to be proved and sealed at the place assigned for that purpose, and shall there prove and seal the same, and shall demand and receive of the owner of every beam, weight, and measure, proved and sealed, two pence, and no more: and every person that shall refuse to have their beams, weights, and measures, viewed, proved, and sealed, shall forfeit the sum of five shillings; one moiety thereof to the use of the poor of the town, and the other moiety to the sealer, to be recovered as is above provided. And if any person shall bring his beam, weights, or measures, to be proved and sealed at any other time than on the day or days set by the sealer for that purpose, he shall in like manner pay two pence for each that shall be tried and sealed.

7. *And be it further enacted by the authority aforesaid,* That if any person, from and after the first day of September next, after the publication of this act, shall sell, vend, or utter, any goods, wares, merchandises, grain, or other commodities whatsoever, by other beams, weights, or measures, than such as shall be proved and sealed as this act requires, the person so offending shall lose or forfeit the sum of five shillings for each offence of that kind; one moiety thereof to the use of the poor of the town where the offence shall be committed, the other moiety to the sealer or informer, who shall prosecute the same, to be heard and determined by one or more of his majesty's justices of the peace.

8. *And be it further enacted by the authority aforesaid,* That all beams, weights, and measures, kept for standards in the several towns, shall be proved and tried by the public standard at the end of ten years, from time to time; and all town standards shall be stamped with this mark, P N H; any law, usage, or custom, to the contrary notwithstanding.

APPENDIX.

E 1—c.

Copy of an Act passed 6th George III.

An Act in addition to an Act, entitled " An Act for regulating Weights and Measures."

Whereas the said act, by experience, is found ineffectual to answer the good end thereby intended, as the penalties therein imposed are insufficient to enforce a due observance thereof: Wherefore,

1. *Be it enacted by his Excellency the Governor, Council, and Representatives, convened in General Assembly, and it is hereby enacted and ordained by the authority of the same,* That, from and after the passing of this act, every person who shall neglect to bring in their weights and measures at the time and place appointed, (being duly warned thereof,) shall forfeit the sum of forty shillings, to be recovered and applied in the same manner as by the said act is directed, for recovering the fine therein inflicted for such neglect.

2. *And it is further enacted by the authority aforesaid,* That when and so often as the sealer of weights and measures, in any town or parish within this province, shall have probable cause of suspicion that any inhabitant has two sets of weights and measures, according to one, whereof, (being legal) the said inhabitant buyeth, and with the other (being lighter or smaller) he selleth, and secreteth the latter, or produceth not the same to the sealer, it shall and may be lawful for the said sealer, verbally, to warn the said inhabitant to appear before the next justice of the peace for the said province, who is hereby authorized and required to examine the said inhabitant upon oath (without fee or reward) touching the same weights and measures, that so the fraud (if any there be) may be detected; and if the said inhabitant, so verbally warned as aforesaid, shall refuse to attend upon the said justice as aforesaid, the said justice, upon satisfactory proof of the said warning, shall issue his warrant to apprehend such delinquent, and bring him, or her, before him, when, if the said delinquent shall refuse to answer upon oath, he, or she, shall incur and forfeit the same penalty as in the said act is inflicted on persons who shall sell, vend, or utter, any goods, wares, merchandise, grain, or other commodities, by other beams, weights, or measures, than such as shall be proved and sealed as the same act requires, and pay cost of prosecution.

3. *And be it further enacted,* That, for the future, the selectmen of the said towns and parishes, (not already provided) shall, at the charge of the towns and parishes, respectively, procure all those weights and measures which, by the law aforesaid, are to be provided by the town clerks, and improved and used as standards for the said towns and parishes; and in default thereof, for the term of six months from the passing of this act, those selectmen who shall be delinquent herein shall forfeit and pay the sum of ten pounds, for the

use of the poor of the town or parish where the selectmen shall be so negligent, to be levied by distress and sale of their goods and chattels, by warrant from any justice of the peace.

E 1—d.

An act regulating scale beams, steelyards, weights, and measures.

SEC. 1. *Be it enacted by the Senate and House of Representatives, in General Court convened,* That the governor, by and with the advice of council, be, and hereby is, authorized and empowered to appoint a sealer of weights and measures in each county in this state.

SEC. 2. *And be it further enacted,* That each sealer of weights and measures, appointed as aforesaid, shall provide, at the expense of the state, one complete set of scale beams, weights, and measures, similar to those now owned by this state; which shall be kept by him as standards for the use of said county. And it shall be the duty of said sealer of weights and measures to try and prove by said standards all scale beams, steelyards, weights, and measures, which shall be brought to him for that purpose by the sealers of weights and measures chosen in the respective towns in said county; and to seal such as shall be found just, agreeable to said standards, who shall receive six cents for every scale beam, steelyard, weight, and measure, so tried, proved, and sealed.

SEC. 3. *And be it further enacted,* That the selectmen of every town in this state shall provide, at the proper expense of their respective towns, one complete set of weights and measures, and a scale beam as aforesaid, for the use of said town, of such materials as the town shall think proper, provided the liquid measures be of some kind of metal.

SEC. 4. *And be it further enacted,* That each town in this state shall, at their annual meeting, choose one suitable person for sealer of weights and measures in said town, who shall be sworn to the faithful discharge of his duty, who shall notify the inhabitants to bring in all scale beams, steelyards, weights, and measures, as they make use of, in the month of May, from year to year, at such time and place as he shall appoint, by posting up a notification at every meeting house in said town, and if there be no meeting house, then at some public place in said town, three weeks successively prior to the day appointed; and the said sealer shall try, prove, and seal, all such scale beams, steelyards, weights, and measures, as shall be brought to him, and shall be found just, agreeable to said standards : and he shall have for every scale beam, steelyard, weight, and measure, so sealed, two cents from the owner thereof at the first sealing, and after the first sealing one cent only, so long as they continue just with the standard.

SEC. 5. *And be it further enacted,* That all measures by which meal, fruit, and other things usually sold by heap, shall be sold, be of the following dimensions, viz : the bushel not less within side than eighteen inches and a half wide; the half bushel not less than thirteen inches and three quarters wide ; the peck not less than ten inches and three quarters wide ; and the half peck not less than nine inches wide. And if any person at any time from and after the first day of September next, shall sell, expose to sale, or offer any meal, fruit, or other things usually sold by heap, by any other measure than is aforementioned, as to bigness and breadth, such person, being complained of, and convicted before any justice of the peace within the county of so doing, shall forfeit and pay to the use of the poor of the town where the offence is committed, the full value of the meal, fruit, or other things so sold, or offered to sale, with costs.

SEC. 6. *And be it further enacted,* That the sealer of weights and measures appointed in each town within this state, from time to time, shall be, and hereby is, empowered to go to the houses of such of the inhabitants, having been duly notified as aforesaid, who shall neglect to bring or send in their scale beams, steelyards, weights, and measures, to be proved and sealed at the place assigned for that purpose, and shall there prove and seal the same, and shall receive of the owner for every scale beam, steelyard, weight, and measure, proved and sealed, twenty cents and no more ; and every person that shall refuse to have their scale beams, steelyards, weights, and measures, viewed, proved, and sealed, shall forfeit the sum of ten dollars, one moiety thereof to the use of the poor of the town, and the other moiety to the sealer ; and if any person shall bring his scale-beams, steelyards, weights, or measures, to be proved and sealed, at any other time than on the day or days set by the sealer of weights and measures for that purpose, he shall, in like manner, pay three cents for every scale-beam, steelyard, weight, or measure, that shall be tried and sealed, and one cent and a half for such as do not need sealing.

SEC. 7. *And be it further enacted,* That if any person, from and after the first day of September next, shall sell, vend, or utter, any goods, wares, merchandises, grain, or other commodities whatsoever, by other scale-beams, steelyards, weights, or measures, than such as shall be proved and sealed as this act requires, in any town where provision is made, and notification given agreeably to this act, or shall fraudulently so sell, utter, or vend, any goods, wares, merchandises, grain, or other commodities, by any scale-beams, steelyards, weights, or measures, that may be so sealed, that shall prove unjust, the person so offending shall forfeit a sum not less than one dollar, nor more than ten dollars, with costs, for each offence ; one moiety thereof to the use of the poor of the town where the offence shall be committed, the other moiety to the informer who shall prosecute the same.

SEC. 8. *And be it further enacted,* That all scale-beams, steelyards, weights, and measures, kept for standards in the several towns, shall be proved and tried by the public county standards at the end of every five years from time to time.

SEC. 9. *And be it further enacted,* That if the selectmen of any town in this state neglect to comply with their duty in procuring weights and measures, and a scale-beam, as by this act is required, they shall forfeit the sum of one hundred dollars, to be recovered, one half for the use of the county in which the neglect shall happen, and the other half for the use of the person who shall sue for the same.

Sec. 10. *And be it further enacted,* That when any sealer of weights and measures, that may be duly appointed in any town where a scale-beam, weights, and measures, are provided according to this act, shall neglect to notify the inhabitants as aforesaid, shall forfeit the sum of fifty dollars, and for neglecting the duties of his office in any other respect, from one to twenty dollars ; one half for the prosecutor, the other half for the use of the town where such neglect shall happen. And all penalties and compensations mentioned in this act, may be sued for and recovered by action, bill, plaint, or information, in any court proper to try the same.

Sec. 11. *And be it further enacted,* That the sealer of weights and measures for each county may make use of such seal as he may think proper, provided a description thereof, in writing, be lodged in the secretary's office before it be made use of, and that the sealer of weights and measures chosen by each town, respectively, shall use such seal as the town may agree on, a record of which being previously made in the town records.

Provided, That this act shall remain in force till superseded by an act of the general government.*

Approved, December 15, 1797.

E 1—e.

Governor of New Hampshire to the Secretary of State.

New Hampshire Executive Department,
Epping, January 4, 1819.

Sir : I regret that circumstances beyond my control have prevented me till this time from returning an answer to that part of your letter of November 4, 1818, requesting a statement of the scale beams, weights, and measures, owned by this state, and where, and in whose custody, they are kept.

The state owns, of *dry measures,* a bushel, a half bushel, and other measures of that kind, down to the smallest denomination. These measures are of copper, the bushel weighing about 100 lbs. and the others in proportion.

Of *liquid measures* there are a gallon, and half gallon ; these are of block tin.

The weights are a 56 lb. 28 lb. and 14 lb. and two or three of a smaller denomination ; these also are of copper.

The measures and weights are at Portsmouth, in the office of Robert Neal, Jr. Esq. commissary general of this state.

---

* The operation of this act was postponed by four several acts till 10th December, 1801.−

As to scale beams and steelyards, I can find none owned by this state.

I have the honor to be, &c.

WILLIAM PLUMER.

Hon. JOHN QUINCY ADAMS,
Secretary of State U. S. Washington.

E 2—a.

VERMONT.

STATE OF VERMONT, EXECUTIVE DEPARTMENT,
Shaftsbury, January 20, 1818.

SIR: In compliance with a request from the Department of State, I have the honor to enclose to you a copy of a law of this state relating to weights and measures, which is the only act now in force in this state relative to a standard of weights and measures.

Mere accident, and not intentional delay, has prevented a more speedy compliance with said request.

I am, sir, with great respect,
Your most obedient servant,
JONAS GALUSHA.

Hon. JOHN Q. ADAMS,
Secretary of State of the U. S.

E 2—b.

An Act relating to Weights and Measures.

SEC. 1. *It is hereby enacted by the General Assembly of the state of Vermont,* That the treasurer of this state shall provide, and keep in good order and repair, in his office, one complete set of weights and measures necessary for the use of this state, according to the approved Winchester measure allowed in England in the exchequer, namely: one half bushel, one peck, one half peck, one ale quart, one wine gallon, one two quart wine measure, one quart, one pint, one half pint, one gill, and one half gill wine measure; one English ell, one yard, one set of iron weights, namely: one fifty-six pound weight, one twenty-eight pound weight, one fourteen, one seven, one four, one two, and one pound weight, and a suitable scale and beam necessary for the use of the same; also, one set of brass weights, from one ounce to four pounds, at the rate of sixteen ounces to the pound, and a suitable scale and steel-beam, necessary for the use of the same, tried and approved according to said standard of Winchester; which shall be considered and understood to be the public standard throughout this state, for the approving and sealing all weights and measures. Which standard of weights and measures shall be provided by the treasurer of this state, from time to time, as the same shall become necessary, and kept in his office. And, if the said treasurer shall neglect to procure and keep in his office aforesaid all or any of the weights, measures, scales, or beams, as aforesaid, he shall forfeit and pay *one hundred dollars* for each and every six months he shall be so deficient, with costs, to be recovered before any court of competent jurisdiction; one moiety of which sum to the prosecutor, and the other moiety to the treasury of the county in which such prosecution shall be had. And each and every county treasurer within this state shall, at the expense of their respective counties, provide, within six months, and keep the same in repair in his office, all the aforesaid weights, measures, beams, and scales, according to the standard abovementioned, proved and sealed by the treasurer of this state; and shall, from time to time, keep the same in his office in good order and repair. And the selectmen of every town within this state shall provide, from time to time, as they may be wanted, at the expense of the town, one half bushel, and one peck, of the following dimensions, namely : the half bushel in diameter, within side, not less than thirteen inches and three quarters of an inch; the peck not less than ten inches and three quarters within side; one half peck, one ale quart, one wine gallon, one two quart, one quart, one pint, one half pint, one gill, one half gill, wine measure; one English yard; one set of brass weights from one ounce to four pounds avoirdupois weight, with scales and steel beam; all the above measures, weights, scales, and beams, to be tried, proved, and sealed, by the county treasurer, according to the aforesaid standard of Winchester provided by the respective counties, which shall be kept only for standards.

Sec. 2. *And it is hereby further enacted,* That all steelyards that are, or shall be, approved of by the standard, shall be allowed of in any town of this state, and be at the liberty of both buyer and seller to weigh by.

Sec. 3. *And it is hereby further enacted,* That all weights, measures, scales, and beams, provided by the respective towns within this state as aforesaid, shall yearly, and every year, be delivered to the sealer of weights and measures, who shall be chosen and sworn, as the law directs, in the several towns. And the sealer of weights and measures, within every town in this state, shall post up a notification in writing, in the month of January annually, requiring all and every person, within their respective towns, to bring in to said sealer of weights and measures all such weights and measures by which they respectively buy or sell, giving, at least, fourteen days' notice of the time appointed for sealing as aforesaid. And the sealers of such weights and measures may demand and receive, from the owner of all weights and measures, so tried, proved, and sealed, by the town seal, two cents for each weight or measure so sealed by him. And, if any person shall carry any weights or measures to be sealed at any time, after the day notified for sealing as aforesaid, the sealers of weights and measures, in such case, may demand and take eight cents for each article sealed; which town seals shall be provided by the selectmen of their respective towns, at the expense and charge of the same.

SEC. 4. *And it is hereby further enacted*, That if the treasurers of the respective counties, or the selectmen or sealers of weights and measures of any town in this state, shall neglect or refuse to procure any weights, measures, scales, or beams, or shall neglect or refuse to seal any weight or measure, or to give notice as above directed, they shall severally forfeit, and pay for every month's neglect or refusal the sum of *three dollars* and *fifty cents*; one moiety to the prosecutor, and the other moiety for the use of the poor of the town in which such delinquent lives, to be recovered by action of debt, before any court proper to try the same.

SEC. 5. *And it is hereby further enacted*, That if any person, or persons, within this state, shall, twenty days after notice given by the sealers of weights and measures as aforesaid, sell or vend any wares, merchandise, or other commodities whatever, by any other beams, weights, and measures, but such as shall be tried, proved, and sealed, as this act requires, the person, so offending, shall forfeit, for each offence, a sum not exceeding *seven dollars*; one moiety thereof to the prosecutor, the other moiety for the use of the poor of the town, to be recovered in manner as is hereinbefore provided.

SEC. 6. *And it is hereby further enacted,* That all beams, weights, and measures, kept for standards in the several towns, shall be tried and proved every ten years by the county standard. And the state standard shall be stamped with the letters S. S., the several county standards shall be stamped with the letters C. S., and the several town standards with the letters P. D.

E 3—a.

MASSACHUSETTS.

Governor of Massachusetts to the Secretary of State.

MEDFORD, *October 24th*, 1817.

SIR: I have the honor to transmit to you, enclosed herewith, an abstract of the laws of the commonwealth of Massachusetts for regulating weights and measures.

With great respect,
I have the honor to be, sir,
Your most obedient servant,
J. BROOKS.

Hon. JOHN QUINCY ADAMS,
Secretary of State.

APPENDIX. 181

E 3—b.

COMMONWEALTH OF MASSACHUSETTS,
Secretary's Department, Sept. 5th, 1817.

To his Excellency Governor Brooks.

SIR: In obedience to instructions from your excellency to furnish an abstract of the laws now in force in this commonwealth respecting *weights and measures,* for the purpose of complying with a request of the acting Secretary of State of the United States, made to your excellency by a note of the 29th of July last, in pursuance of a resolution of Congress of March 3d, 1817, I have now the honor to submit the following statement on that subject.

In February, 1800, an act passed the legislature of this commonwealth, entitled "An act for the due regulation of weights and measures," by which it was required that the brass and copper weights and measures formerly sent out from England with a certificate from the exchequer, to be approved Winchester measures, according to the standard in the said exchequer, and adopted and used, and allowed in this commonwealth, should be and remain the public allowed standards through the same, by which all 'weights and measures should be tried, proved, and sealed.

By said act it was made the duty of the treasurer of the commonwealth to have and keep, as public standards, to be used only as such, the *beams, weights,* and *measures* following, to wit: one bushel, one half bushel, one peck, one half peck, one ale quart, one wine gallon, one wine half gallon, one wine quart, one wine pint, one wine half pint, and one wine gill; the said measures to be of copper or pewter, conformable as to contents to said Winchester measures, and as to breadth, that is to say, the diameter of the bushel not less than eighteen inches and a half, containing thirty-two Winchester quarts; of the half bushel, not less than thirteen inches and three quarters, consisting of sixteen Winchester quarts; of the peck, not less than ten inches and three quarters, containing eight Winchester quarts; and of the half peck, not less than nine inches, containing four Winchester quarts; the admeasurement to be made in each instance within side the measure. Also, one ell, one yard, one set of brass weights to four pounds, computed at sixteen ounces to the pound, with fit scales and steel-beam. Also, a good beam and scales, and a nest of troy weights, from one hundred and twenty-eight ounces down to the least denomination, with the weight of each weight, and the length of each measure, marked or stamped thereon, respectively, and sealed with a seal, to be procured and kept by the said treasurer of the commonwealth. And, also, one fifty-six pound weight, one twenty-eight pound weight, one fourteen pound weight, and one seven pound weight, made of iron.

APPENDIX.

By the same act the county treasurers were also required to keep a complete set of beams, and of brass, copper, pewter, and iron weights, and of the aforesaid measures, (the bushel measure excepted) tried, proved, and sealed, by the said state standards, and marked or stamped as aforesaid, the measures to be conformable, as to breadth and contents, to the state standards, to be kept for the sole use of the respective counties, and to be used only as standards, and every ten years to be tried and proved by the treasurer and standards of the commonwealth; and, for neglect of duty in this behalf, the county treasurers are made liable to a fine of two hundred dollars.

The treasurers of towns were also required by the same law to procure and keep for town standards a complete set of *beams*, weights, and copper or pewter measures, under a penalty of one hundred dollars, to be conformable to said state standards, (excepting as to a bushel measure, and with liberty also to have a wooden, instead of copper, iron, or pewter half bushel, peck, and half peck.) The town treasurers also to be excused from procuring a nest of troy weights, other than from the lowest denominations to the size of eight ounces; the same to be tried, proved, and sealed, either by the treasurer of the commonwealth, or of the county within which the town lies, and to be proved and sealed every ten years. Said town treasurer also to procure and keep a town seal for the purpose of sealing weights, &c. And the selectmen of every town are directed by the same law, annually, to appoint a suitable person as sealer of weights and measures, with power to remove and appoint others. The person appointed to be sealer of weights and measures to be under oath; the selectmen are liable to a fine of ten dollars for neglect of duty in this case, and the sealer of weights and measures to a fine of one hundred dollars for neglect of duty. The sealer to receive the town seals and standards for the purpose of proving, marking, and sealing weights and measures.

The sealer of weights and measures is also required by said law to give public notice of his appointment, and of the time when, and place where, he will try, prove, and seal the measures and weights. And he is authorized to go to the houses, stores, or shops, of such as do not come to him to have their measures and weights proved, &c. And those who refuse to have beams, weights, and measures so tried, proved, and sealed, to forfeit ten dollars for every offence. Those who offer to sell, or do sell, without such proved and sealed weights and measures, to forfeit one dollar for every offence.

Fees to be paid state or county treasurer for sealing any weight, measure, scale, or beam, the first time three cents, and two cents for every after sealing of the same. The sealer of weights and measures for trying and proving town standards, and sealing them, to receive three cents, if they are found not conformable to the standard, and one cent and five mills for each beam, weight, and measure, if conformable thereto.

APPENDIX.

In 1804, March, an act passed the legislature, in addition to the act above referred to, and abridged, requiring the treasurer of the commonwealth to add to the troy weights, (by the first act ordered to be kept by him for public use,) as follows, viz: To the weight of one hundred and twenty-eight ounces, the further weight of twenty-seven grains; to the weight of sixty-four ounces, the further weight of fifteen grains; to the weight of thirty-two ounces, the further weight of six grains; to the weight of sixteen ounces, the further weight of seven grains; to the weight of eight ounces, the further weight of four and a half grains; to the weight of four ounces, the further weight of two and a half grains; to the weight of two ounces, the further weight of two and a half grains; to the weight of one ounce, the further weight of two grains; to the weight of half an ounce, the further weight of one quarter of a grain; or to procure new weights of the same denomination, and conformable to the state standards, with the additions aforesaid, respectively; and the same to be the standards of troy weight for the commonwealth.

By this last named act, the directors of all banks in the state are required annually to have all the weights used in their respective banks, compared, proved, and sealed, by the treasurer of the commonwealth, or by some person specially appointed by him for the purpose; and the weights of banks are not, therefore, required to be sealed by town treasurers or sealers. And no tender of gold by any bank is to be legal, unless weighed with weights thus compared, proved, and sealed.

The county treasurers are required by this law to have their standards of troy weights compared, proved, and sealed, by the treasurer of the commonwealth, once in ten years, at the expense of the county. And the treasurers of towns, at the expense of the towns, to have the town standards of troy weight compared, proved, and sealed, every ten years, by the treasurer of the commonwealth, or the treasurer of the county.

The above is the substance of the laws of this state now in force through the whole commonwealth on the subject of weights and measures. In June, 1817, a law passed regulating weights and measures in the town of Boston. The provisions of this law vary but little from the former law above abridged, which is obligatory through the state. It provides that the sealer of weights and measures for the town of Boston shall have a house or office, to which all persons in Boston, using scale-beams, steelyards, weights, or measures, for the purpose of buying or selling any article, shall be obliged, after due notice in the public papers, to send annually their scale-beams, steelyards, weights, and measures, to be tried, proved, and sealed, as required by the act first above referred to. The sealer in Boston, by this act, is authorized and required to go to the houses or shops of those who do not send their weights, scales, &c. to him, to try, prove, and seal the same; and is to have double fees in such case. For refusing to have scales, weights, &c. proved and sealed, and for using weights, scales, &c. which are not conformable to the state standards, a fine of ten dollars is provided for each offence. And for altering any scale or weight, &c. a penalty of fifty dollars for each offence.

ALDEN BRADFORD,
Secretary of Commonwealth of Massachusetts.

E 4.

RHODE ISLAND.

Governor of Rhode Island to the Secretary of State.

THE STATE OF RHODE ISLAND,
Providence, September 5, 1817.

SIR: Agreeably to your request of the 29th of July last, relative to the regulations and standards for weights and measures here, I have to observe there is not any statute of this state regulating weights and measures, though the general assembly have passed some acts regulating the assize of casks, for cider and stone lime, but have never declared the cubical contents of a gallon or bushel.

The weights and measures now in use are sealed by the standard weights and measures procured from England many years since, and are said to be in conformity with those called the *Lower Standards*. However, they are the same as those used in the states of Connecticut and Massachusetts, which are generally known to be the same as the Winchester measure.

I have not been able to find, in the course of my researches, any thing written on the subject of weights and measures by any authority of this state, but I find it the immemorial custom of merchants to buy and sell sugar, rice, hay, iron, hemp, cordage, copperas, and dye-woods, by gross hundreds, that is, 112 pounds for the hundred.

Respectfully, sir,
Your obedient servant,
N. R. KNIGHT.

N. B. Land and distances are measured here by the English rod.

Hon. RICHARD RUSH,
   Acting Secretary of State.

APPENDIX.

E 5—a.

CONNECTICUT.

> STATE OF CONNECTICUT,
> *Litchfield, August 5, 1817.*

SIR: I have received your letter of the 29th of July, and have directed the Secretary of State to transmit, to your department, exemplifications of the acts of this state, establishing the standards of weights and measures.

> I have the honor to be,
> With perfect respect, sir,
> Your most obedient servant,
> OLIVER WOLCOTT.

The Hon. RICHARD RUSH,
Acting Secretary of State, Washington.

E 5—b.

> STATE OF CONNECTICUT,
> *Secretary's Office, August 12, 1817.*

SIR: In compliance with the request of his excellency the Governor of this state, I enclose, herewith, for your use, exemplifications of all the statutes of this state now in force relating to weights and measures.

> With great respect,
> I have the honor to be,
> Your obedient servant,
> THOMAS DAY.

The Hon. RICHARD RUSH,
Acting Secretary of State, Washington.

E 5—c.

STATE OF CONNECTICUT.

At a general assembly of the state of Connecticut, in America, holden at New Haven, in said state, on the second Thursday of October, being the ninth day of said month, and continued by adjournments from day to day, until the first day of November, in the year of our Lord one thousand eight hundred.

An Act prescribing and regulating Weights and Measures.

SEC. 1. *Be it enacted by the Governor and Council, and House of Representatives, in General Court assembled,* That the brass measures, the property of this state, kept at the treasury, that is to say: a half bushel measure containing one thousand and ninety-nine cubic inches, very near, a peck measure, and a half peck measure, when reduced to a just proportion, be the standard of the corn measures in this state, which are called by those names respectively; that the brass vessels ordered to be provided by this assembly, one of the capacity of two hundred and twenty-four cubic inches, and the other of the capacity of two hundred and eighty-two cubic inches, shall be, when procured, the first of them the standard of a wine gallon, and the other the standard of ale or beer gallon in this state; that the iron or brass rod, or plate, ordered by this assembly to be provided, of one yard in length, to be divided into three equal parts for feet in length, and one of those parts to be subdivided into twelve equal parts for inches, shall be the standard of those measures, respectively; and that the brass weights, the property of this state, kept at the treasury, of one, two, four, seven, fourteen, twenty-eight, and fifty-six pounds, shall be the standard of avoirdupois weight in this state.

SEC. 2. *Be it further enacted,* That the treasurer of this state, for the time being, shall have the custody and safe-keeping of all the aforesaid weights and measures; and it shall be his duty, personally, or by some meet person or persons by him appointed, to try all such weights and measures as shall, pursuant to the provisions of this act, be presented to him to be tried by the proper standard, and to seal such as are found true with the capital letters S. C.

SEC. 3. *Be it further enacted,* That the treasurer of each county shall, on or before the first day of January next, provide, and constantly keep and preserve in good order, weights and measures correspondent to all the aforesaid standards, and of like materials, and shall, within the time aforesaid, cause them to be tried and sealed by those standards; and, in default thereof, shall, on conviction, before the county court of the same county, pay a fine of seventeen dollars to the treasury of such county, to be recovered at the suit of the state's attorney for such county; and the county treasurer, for the time being, after such conviction, shall incur a like penalty every term of three months he shall neglect his duty herein prescribed, to be recovered as aforesaid, for the use aforesaid: *Always provided,* That the state standards aforesaid may be used and improved, as heretofore, as standards for weights and measures for the county of Hartford.

SEC. 4. *Be it further enacted,* That the treasurer of each county, for the time being, shall have the custody and safe-keeping of the weights and measures belonging to the county, and it shall be his duty, either personally, or by some meet person or persons by him deputed, to try all such weights and measures as shall, pursuant to the provisions of this act, be presented to him to be tried by the county standard, and to seal such as are found true with the capital letter C, and that which begins the name of the county.

SEC. 5. *Be it further enacted,* That the selectmen of each town shall, on or before the first day of March next, at the cost and charge of the town, provide weights and measures of the various kinds afore-

APPENDIX. 187

said, of good and sufficient materials, which, for the standards of liquid measure, shall be copper, brass, or pewter, as standards for such town, and cause the same to be tried and sealed by the county standards; also, the following vessels for corn measure, of the forms and dimensions herein described, to wit: a two quart measure, the bottom of which on the inside is four inches wide on two opposite sides, and four inches and a half on the two other sides, and its height from thence seven inches and sixty-three hundredths of an inch; a quart measure, the capacity of which is three inches square from bottom to top throughout, and its height seven inches and sixty-three hundredths of an inch; and a pint measure, the capacity of which from bottom to top is three inches square throughout, and its height three inches and eighty-two hundredths of an inch, and, in default thereof, such selectmen shall, on conviction before an assistant or justice of the peace, forfeit and pay a fine of seven dollars; the one half to him or them who shall prosecute to effect, and the other half to the town treasury. And all informing officers are required to inquire after, and due presentment make of, all breaches of this act, and, after such conviction, the selectmen, for the time being, shall incur a like penalty for every term of two months they shall neglect their duty herein prescribed, or shall, at any future time, fail to preserve such weights and measures, true and in good order for the use of the town, to be recovered as aforesaid for the use aforesaid.

SEC. 6. *Be it further enacted*, That the sealers of weights and measures in each town shall have the custody and safe-keeping of the weights and measures belonging to the town, respectively, and it shall be their duty, once in every year, to try the several weights, steelyards, and measures, that are used and improved by any person, or persons, in such town, by the town standards; to deface and destroy all such as cannot be brought equal with the standard, and to seal with the capital letter initial in the name of the town all such as are found or made true; and such sealer shall, sometime in the month of April yearly, give notice in writing to the inhabitants of the town, posted on the sign post and other public places in the town, to bring their steelyards, weights, and measures, at time and place therein fixed, to be tried and sealed. And if any person, or persons, shall, after the first day of May next, for the purpose of buying or selling, use any weight or measure until the same shall have been sealed in the manner aforesaid, each person, so offending, shall, for every such offence, forfeit the sum of two dollars, one half for the benefit of the town in which such offence shall be committed, and the other half for the benefit of the sealer of weights and measures for said town, whose duty it shall be to prosecute the same to effect.

SEC. 7. *Be it further enacted*, That every sealer who shall neglect his duty required in this act, shall forfeit the sum of five dollars for every such neglect, to the town treasury.

SEC. 8. *Be it further enacted*, That the statute entitled "An act for due regulation of weights and measures," be, and the same is hereby, repealed.

APPENDIX.

E 5—d.

STATE OF CONNECTICUT.

At a general assembly of the state of Connecticut, in America, holden at New Haven, in said state, on the second Thursday of October, being ——— day of said month, and continued by adjournments from day to day, until the ——————— in the year of our Lord one thousand eight hundred and one.

An act in alteration of the statute, entitled " An act prescribing and regulating weights and measures."

SEC. 1. *Be it enacted by the Governor and Council and House of Representatives, in General Court assembled,* That the treasurer of this state do, without delay, provide a vessel of brass, the capacity of which shall be five inches square from bottom to top throughout, and nine inches and twenty-four hundredths of an inch in height, containing two hundred and thirty-one cubic inches, and the same, when provided, shall be the standard of a wine gallon in this state; any thing in said act to the contrary notwithstanding.

SEC. 2. *Be it further enacted,* That the provisions of said act shall relate to the said standard in all respects as they related to that in place of which it is substituted: and the treasurer of each county, by the first day of January next, and the selectmen of each town, by the first day of March next, shall, respectively, on penalty as by said act is in each case provided, to be recovered and applied as therein directed, provide, and cause to be tried and sealed by the proper standard, and constantly kept in good order, a vessel or measure of the same capacity as and for a standard of a wine gallon, for such counties and towns respectively.

At a general assembly of the state of Connecticut, holden at Hartford, in said state, on the second Thursday of May, in the year of our Lord one thousand eight hundred and ten.

An act regulating the measure of charcoal.

Be it enacted by the Governor and Council, and House of Representatives, in General Court assembled, That the standard measure of charcoal shall be the half bushel measure prescribed by the act, entitled " An act prescribing and regulating weights and measures," and that in measuring charcoal such measure shall be well heaped.

General Assembly, May session, 1810.

LYMAN LAW,
Speaker of the House of Representatives.
JOHN TREADWELL,
Governor.

Attest—THOMAS DAY, *Secretary.*

APPENDIX.

E 6—*a*.

NEW YORK.

The Governor of New York to the Secretary of State.

ALBANY, 4*th September*, 1817.

SIR: In compliance with the request contained in your letter of the 29th of July, I now transmit a copy of the law of this state relative to weights and measures, marked number 6. The papers numbered 1, 2, 3, 4, and 5, exhibit our regulations on this subject, at different periods.

I have the honor to be, very respectfully,
Your most obedient servant,
DE WITT CLINTON.

The Hon. R. RUSH.

E. 6. *b*.

An act to ascertain the assize of casks, weights, measures, and bricks, within this colony. Passed June 19, 1703.

Whereas nothing is more agreeable to common justice and equity, nor for the good and benefit of any people or government, who live in community and friendship together, than that they have one equal and just weight and balance, one true and perfect standard and assize of measure among them; for want whereof experience shews that many frauds and deceits happen, which usually fall heavy upon the meanest and most indigent sort of people, who are least able to bear the same, and may be accounted little better than oppression. For remedy of which evil,

Be it enacted, and it is hereby enacted, by his excellency the Governor, by and with the consent of her majesty's council and representatives of this colony, &c. That, from and after the first day of August next, no cooper, or other person or persons whatsoever, within this city or colony, shall make any dry cask or vessel but of good and well seasoned timber, and of the respective dimensions following, viz.: Every hogshead to be forty inches long, thirty-three inches in the bulge, and twenty-seven inches in the head. Every tierce to be thirty-six inches long, twenty-seven inches in the bulge, and twenty-three inches in the head. Every barrel to be thirty inches long, twenty-six inches in the bulge, and twenty-two inches in the head. Every half barrel to be twenty five inches long, twenty inches in the bulge, and sixteen inches in the head. Every quarter barrel to be twenty inches long, sixteen inches in the bulge, and thirteen inches in the head. All tight barrels to contain thirty-one gallons and a half of wine measure each, and not to exceed or be half a gallon over or under the same, and all other casks to contain in proportion to a barrel, upon penalty of five shillings for every offence committed to the contrary hereof, to be paid by the maker or user of such casks or vessels, in whose hands the said offence shall first happen to be known or discovered.

And to the preventing other frauds and deceits that may be in casks made as aforesaid,

Be it further enacted, &c. That, from and after the said first day of August, all and every the cask and casks, which shall be employed or used for the stowing or packing of flour or biscuit, within this city or province, for transportation thereof, or otherwise, in any way, of merchandise, before any of the said goods or commodities shall be put or packed therein, shall be truly weighed, and the just weight and tare thereof be set with a marking iron upon the head of each cask so employed as aforesaid, together with the name of each respective person using or employing the same, upon the penalty of nine pence, to be paid for every neglect herein, by the person or persons, respectively, on whose account any of the goods or commodities aforesaid shall, to the contrary hereof, be stowed or packed as aforesaid.

And be it further enacted, &c. That, from and after the first day of August aforesaid, there shall be one just beam or balance, one certain weight and measure, and one yard, that is to say, avoirdupois and troy weights, bushels, half bushels, pecks, and half pecks, according to the standard of her majesty's exchequer, in her realm of England, throughout all this colony, as well in places privileged as without, any usage or custom to the contrary notwithstanding; and that every measure of corn shall be striked without heap. And whosoever sell, buy, or keep any other beam, weight, measure, or yard, than as aforesaid, whereby any corn, grain, or other thing, is bought or sold, from and after the time limited as aforesaid, shall forfeit for every such offence 20s.

And for the better observance and putting in execution of this act, fit persons be appointed in all counties and cities within this colony, for the sealing and marking all beams, weights, measures, and yards, to be used within the respective counties and cities aforesaid, with the letter A, according to the standard of her majesty's exchequer in England, that the same may be known throughout this colony; and that his excellency the governor aforesaid be desired to nominate and appoint such fit persons in all proper places within this colony aforesaid, the which respective persons, when nominated and appointed, shall take for their pains in sealing and marking all such beams, weights, measures, and yards, as shall from time to time for that purpose be brought in to them, the rate of nine pence, except weights and small liquid measures, which shall pay only one penny each, and no more, on penalty of five shillings for the least exaction therein, saving always, nevertheless, unto the cities of New York, Albany, and borough of Westchester, and the mayors thereof, for the time be-

APPENDIX. 191.

ing, all such rights, privileges, and usages, as they respectively can justly claim, as clerks of the market within the said cities and borough, or otherwise howsoever, any thing herein contained to the contrary hereof in any wise notwithstanding.

And be it further enacted, &c. That, from and after the first day of August, no person or persons, be he master or servant, shall make, or suffer to be made, in any place or places within this colony, any bricks, or kiln of bricks, but such as shall be well and thoroughly burnt, and of the size and dimensions following, that is to say : every brick to be and contain nine inches in length, four inches and one quarter of an inch in breadth, and two inches and one half inch in thickness thereof, all well struck off in good and workman-like order and manner, and made of right and well tempered mould, or clay, on penalty of six shillings for every neglect herein, to be paid by the master or owner of the said brick, or kiln, in whose hands or wheresoever the neglect or offence aforesaid shall be discovered or found out, except well bricks.

And it is hereby also further enacted, &c. That, from and after the time limited as aforesaid, no other casks, beams, weights, measures, yards, or bricks, shall be used within this colony than such as aforesaid, except well bricks and such other bricks as are already made or to be made before the commencement of this act, on the penalty of twenty shillings, to be paid by the person using the same, or any of them.

Provided always, That all and every the penalties and forfeitures in and by this act set and appointed as aforesaid, shall be one half to the use of the poor of the parish, town, or place, where the default or offence happens to be; the other moiety thereof to the use of the person or persons who shall inform and sue for the same forfeitures in any of her majesty's courts of record within this colony, or else to be recovered to the uses aforesaid, upon conviction of the offender by the oath of one sufficient witness before any justice of the peace, mayor, or other head officer of the city, county town, or place, respectively, where the offence against, or breach of this act, shall be committed, (who, by virtue of this act, shall have power to administer an oath on that behalf,) by way of distress and sale of the offender's goods and chattels; the overplus, if any be, after charges of the distress deducted, to be returned to the owner thereof, and where no distress can be had, that it shall and may be lawful to and for any justice of the peace, mayor, or head officer aforesaid, to commit the said offender or offenders to prison, there to remain without bail or mainprize until he or they shall pay the penalties and forfeitures aforesaid for which they shall be so committed, to the uses aforesaid. *Provided, also,* That no prosecutions shall be for any of the forfeitures aforesaid but within three months after the respective facts are committed; any thing herein to the contrary hereof notwithstanding.

E 6—c.

An act to ascertain weights and measures within this state. **Passed** 10*th April,* 1784.

Whereas it is agreeable to equity, and beneficial to commerce, that a people who live in the same community shall have one equal and just weight and balance, according to a true and perfect standard and assize of measures, to be established by law, without which necessary provision frauds and deceits may be practised with impunity:

1. *Be it therefore enacted by the people of the state of New York, represented in Senate and Asembly, and it is hereby enacted by the authority of the same,* That, from and after the first day of June next, there shall be one just beam, one certain weight and measure; that is to say, avoirdupois and troy weights, bushels, half bushels, pecks, and half pecks, according to the standard in use in this state, on the day of the declaration of the independence thereof: and that the standard weights and measures in the custody of William Hardenbrook, who, before and at the time of the said declaration, was the public sealer and marker of all beams, and weights, and measures, within the city and county of New York, which standard is according to the standard of the court of exchequer in that part of Great Britain called England, shall forever hereafter be deposited with, kept, and preserved by, the clerk of the peace, or common clerk of the city and county of New York, for the time being, and shall be and hereby are declared and established to be and remain the standard, for ascertaining all beams, weights, and measures, throughout the state, any usage or custom to the contrary thereof notwithstanding; and the said clerk of the peace, or common clerk, now, and for the time being, shall take an oath, to be administered to him in open court before the mayor, recorder, and aldermen, of the said city, well and faithfully to preserve the said weights, seals, and measures, and to suffer no other person to make use of the same, except a sworn public sealer and marker of weights and measures: *Provided always,* That the said William Hardenbrook shall deliver the said beam, weights, and measures, to the clerk of the peace, or common clerk, of the said city and county, in the presence of the mayor, recorder, and one or more of the aldermen of the said city, and shall declare, on his solemn oath, that the said beam, weights, or measures, are the same which he received from the court of exchequer aforesaid, according to the best of his knowledge and belief.

2. *Provided always, and be it further enacted,* That if any of the said standard beams, weights, and measures, shall be broken, impaired, or missing, that it shall and may be lawful to and for the mayor and aldermen of the city of New York, in common council convened, to cause to be delivered to the said clerk of the peace, or common clerk, for the time being, any standarded beam, weights, and measures, respectively, to supply such deficiency, taking care that the same is according to the standard established in the late colony,

now state, of New York, immediately preceding the declaration of independence of this state.

3. *And be it further enacted by the authority aforesaid,* That, for the better observance and execution of this act, it shall and may be lawful to and for his excellency the governor of this state for the time being, by and with the advice and consent of the council of appointment, to appoint fit persons in all convenient and proper places within this state, for sealing and marking all beams, weights, and measures: that the persons so to be appointed shall impress, with the letter A, all beams, weights, and measures, to be sealed and marked by each of them, respectively, and shall respectively take and subscribe an oath before one of the judges of the court of common pleas of the county in which he, or they, shall reside, for the faithful execution of the trust to be committed to them by virtue of this act; and the judge before whom such oath shall be taken, shall cause a certificate thereof to be filed with the clerk of the county wherein such judge shall reside: and every such sworn public sealer and marker of weights, seals, and measures, shall be entitled to receive for his pains in sealing and marking all such beams and measures as shall, from time to time, for that purpose be brought to him, the rate of nine pence, and for every weight, and every small liquid measure, one penny, and no more: saving always, nevertheless, unto the cities of New York and Albany, and borough of West Chester, and the mayors thereof for the time being, all such rights, privileges, and usages, as they respectively can justly claim, as clerks of the markets within the said cities and boroughs, or otherwise howsoever; any thing herein contained to the contrary hereof notwithstanding.

E 6—d.

Extract from an act supplementary to the act, entitled " An act to prevent the exportation of unmerchantable flour, and the false taring of bread and flour casks." Passed 7th March, 1788.

" 7. *And be it further enacted by the authority aforesaid,* That the standard weight of wheat brought to the city of New York for sale, shall be sixty pounds nett to the bushel; and in all cases of sales of wheat in the said city by the bushel, if the same shall exceed the standard weight, the buyer shall pay a proportionably greater price; and if the same shall be less than the said standard the buyer shall pay a proportionably less price: *Provided,* That this regulation shall not extend to any special contracts respecting the sales of wheat, whatever may be the weight thereof."

E 6—e.

An act to amend an act, entitled " An act to ascertain weights and measures within this state." Passed February 2, 1804.

Be it enacted by the People of the state of New York represented in Senate and Assembly, That the secretary of this state for the time being shall, ex-officio, be the state sealer of weights and measures; and that there shall be one assistant state sealer, to be appointed from time to time, as occasion may require, by the person administering the government of this state, by and with the advice and consent of the council of appointment, in the county of Oneida; and that there shall be one sealer of weights and measures in each county, the western district of this state, to be appointed by the board of supervisors that shall think proper to appoint one at their annual meeting in October, and one town sealer of weights and measures in each town that shall think proper to elect one, to be elected as other town officers are directed to be elected.

And be it further enacted, That the several state standard weights and measures shall be made of iron, brass, or copper, as the secretary shall direct; and the several county standard weights and measures shall be of such materials as the several boards of supervisors shall direct; and the several town standard weights and measures shall be of such materials as the supervisor of each town shall direct.

And be it further enacted, That it shall be the duty of the said secretary, within nine months after the passing of this act, at the cost of this state, to procure two sets of standard weights and measures, of the same weight and capacity as is mentioned in the act hereby amended, with such beams as he shall think necessary, one set to be kept by himself as a principal state standard, and the other three sets to be delivered to the said assistant state sealers.

And be it further enacted, That the said county sealers of weights and measures shall, at the cost of the respective counties for which they are elected, within six months after being notified of their several appointments by the clerks of the several boards of supervisors, whose duty it shall be to give such notice, to procure a complete set of standard weights and measures, for the use of each respective county; and that the several town sealers of weights and measures, at the cost of the respective towns, shall, within six months after their respective appointments, procure a set of standard weights and measures for the use of the town or towns for which they shall be respectively appointed.

And be it further enacted, That the letters N Y shall be impressed on all the state standard weights, measures, and beams, and on the several county standard weights, measures, and beams, with such other device in addition as the said secretary shall direct for each county, which device shall be recorded in the records of this state, and a copy thereof delivered by the secretary to the assistant state sealer; and the several town standard weights, measures, and beams,

APPENDIX.

shall be impressed by the county sealer in which such town shall be situate, with such other device in addition to the county device, as the board of supervisors shall direct for the several towns, in their several counties: which several town devices shall be recorded by the clerks of the several boards of supervisors, in a book suitable for that purpose, and a copy of such record delivered by said clerk to the county sealer : *Provided, nevertheless,* That if any town shall neglect to appoint a town sealer of weights and measures, then it shall be lawful for the county sealer to seal all the weights, measures, and beams, belonging to the inhabitants of such town so neglecting.

And be it further enacted, That it shall be the duty of the assistant state sealers to have their standard weights and measures compared with the principal state standard once in fourteen years; and that the several county sealers shall have their standard weights and measures compared with one of the state standards once in seven years, and oftener if necessary; and the several town sealers shall have their town standards compared with the county standard once in three years; that before either of the sealers of weights and measures who shall be appointed by virtue of this act shall enter on the duties of his office, he shall take and subscribe an oath or affirmation before one of the judges of the supreme court, court of common pleas, or justice of the peace, of the county wherein such sealer is resident, well and truly, according to the best of his skill and ability, to perform the duties by this act enjoined, and cause a certificate thereof to be filed in the secretary's office, or the office of county or town clerk, as the case may be.

And be it further enacted, That each of the sealers of weights and measures within this state shall be entitled to receive for his service for sealing and marking beams or measures, which shall, from time to time, be brought to him for that purpose, twelve and a half cents, and for every weight, and every small liquid measure, two cents over and above his cost of making them right, if they are not so when brought to him for that purpose.

And be it further enacted, That so much of the act hereby amended as respects the appointment of sealers of weights, measures, fees, and device, shall be, and hereby is repealed, so far as respects the said western district.

E 6—*f.*

An act relative to a standard of long measure, and for other purposes.
Passed *March* 24, 1809.

Whereas the corporation of the city of New York did, in the year one thousand eight hundred and three, procure a brass yard measure, engraved and sealed at the exchequer in Great Britain, and have presented the same to this state, which has been deposited, together with the documents authenticating it, in the secretary's office—Therefore,

Be it it enacted by the people of the state of New York, represented in Senate and Assembly, That the said brass measure is, and shall be, the standard yard measure of this state.

And be it further enacted, That as soon as may be, after the passing of this act, it shall and may be lawful, and it is hereby made the duty of the secretary of this state, to cause the said standard measure to be engraved with the words *State of New York* thereon; and that he cause to be made and procured two brass yard measures for the city and county of New York, engraved with the words *City and county of New York;* one for the assistant state sealer in the village of Utica, engraved with the words, *State of New York;* and one for each of the respective counties of this state, engraved with the name of the proper county, similar to the aforesaid standard, and that be cause the measure, so procured and marked for each county, delivered to the clerk thereof, who shall keep the same for the use of the county, as the standard yard measure of this state for such county, and compare therewith all yard measures or rods, which may be presented to him for that purpose.

And be it further enacted, That the assistant state sealer, and clerk of each county, shall be paid twelve and a half cents for comparing each yard-stick or rod that shall be presented to him for that purpose, over and above the expense of making such stick or rod exactly compare and agree with the said standard measure of the county, when so presented.

And be it further enacted, That no surveyor, for any survey made for one year from the passing of this act, shall give evidence as a surveyor in any court, or elsewhere in this state, in any cause wherein lands is in dispute, respecting the survey or measurement thereof, unless such surveyor shall make oath, if required, that the chain or measure used by him, when surveying or measuring such land, was conformable to the standard measure of this state.

And be it further enacted, That the treasurer, on the warrant of the comptroller, shall pay to the secretary the expenses incurred by virtue of this act.

And be it further enacted, That the act, entitled "An act to direct the secretary to procure a state standard of long measure, and for other purposes," passed the 11th day of April, 1808, be, and the same is hereby, repealed.

And be it further enacted, That it shall and may be lawful for the inhabitants of any town in this state, for their convenience, and by a vote at their annual town meeting, to direct the clerk of such town to procure, and deposit in his office, a standard brass yard, to be sealed by the person authorized to seal and compare such yard, and to be considered as the true yard for all the purposes aforesaid.

APPENDIX.

E 6—g.

An act to regulate weights and measures. Passed March 19, 1813.

1. *Be it enacted by the People of the state of New York, represented in Senate and Assembly.* That there shall be one just beam, one certain weight and measure, for distance and capacity, that is to say: avoirdupois and troy weights, bushels, half bushels, pecks, half pecks, and quarts, and gallons, half gallons, quarts, pints, and gills, and one certain rod for long measure, according to the standard in use in this state, on the day of the declaration of the independence thereof: and that the standard of weights and measures, now in the office of the secretary of this state, which is according to the standard of the court of exchequer in that part of Great Britain called England, shall be, and is hereby declared to be and remain, the standard for ascertaining all beams, weights, and measures, throughout this state, until the Congress of the United States shall establish the standard of weights and measures for the United States.

2. *And be it further enacted,* That the secretary of this state for the time being shall, ex officio, be the state sealer of weights and measures: and that there shall be three assistant state sealers, to be appointed from time to time as occasion may require, by the person administering the government of this state, by and with the advice and consent of the council of appointment, to continue in office during the pleasure of the said council; one of which assistants shall reside in the city of New York, one in the city of Albany, and one in the county of Oneida: and that there shall be one county sealer of weights and measures, in each county in this state, to be appointed by the board of supervisors of the respective counties, at their annual meeting in October, to continue in office during pleasure; and one town sealer of weights and measures in each town in this state, to be elected at the annual town meetings, who shall continue in office for one year, and until another shall be chosen in his stead.

3. *And be it further enacted,* That it shall be the duty of the secretary of this state, within nine months after the passing of this act, in addition to the weights and measures already provided by law, and now remaining with the said secretary, and the assistant state sealer in the county of Oneida, to procure, at the expense of this state, so many weights, measures, and beams, as shall make out four complete standards of weights and measures, both of liquid and dry measures, and avoirdupois and troy weights, with proper beams, and standards, brass rods of long measure; one complete set to be retained in his office, as a principal state standard, and one other set of the said standards to be delivered to each of the assistant state sealers, taking their receipts respectively therefor; and the comptroller is hereby directed to audit the account of the secretary for his expenses in procuring the said additional standards of weights, measures, and beams, and

draw his warrant for the amount on the treasurer, who is hereby directed to pay the same out of any moneys in the treasury not otherwise appropriated.

4. *And be it further enacted,* That the several state standards of weights, beams, and measures, shall be made of iron, brass, or copper, as the secretary shall direct; and the several county standard weights and measures shall be made of such materials as the several boards of supervisors shall direct: and the several town standard weights and measures shall be made of such materials as the supervisors of each town shall direct.

5. *And be it further enacted,* That the said county sealers of weights and measures, shall, at the expense of the respective counties for which they are elected, within six months after being notified of their respective appointments by the clerks of the several boards of supervisors, whose duty it shall be to give such notice, and after receiving from their respective county treasurers, by order of the said board, so much money as may be necessary for the purpose, procure a complete set of the said standard weights and measures for the use of their respective counties: and every such county sealer shall forthwith, after having procured such standard, deliver to the clerk of the board of supervisors a statement, in writing, of the expense thereof, and that such standard is in his possession, and that the several town sealers of weights and measures shall, at the expense of the respective towns, within six months after their appointments, and after having received sufficient money for the purpose, procure a complete set of the said standard weights and measures for the use of the respective towns: and, having procured the same, shall deliver to the clerk of the town, to be filed in his office, such statement, in writing, as is before specified.

6. *And be it further enacted,* That the letters N. Y. shall be impressed on all the state standard weights, measures, and beams, and on the several county standard weights, measures, and beams, such other device in addition as the said secretary shall direct for each county: which device shall be recorded in the secretary's office, and a copy thereof delivered by the secretary to each of the assistant state sealers, and the several town standard weights, measures, and beams, shall be impressed by the county sealer in which such town shall be situate, with such other device, in addition to the state and county device, as the board of supervisors shall direct for the several towns in their respective counties: which several town devices shall be recorded by the clerks of the several boards of supervisors, in a book to be kept for that purpose: and that such clerk shall deliver a copy of such record to every county sealer.

7. *And be it further enacted,* That it shall be the duty of the assistant state sealers to compare their standard weights and measures with the principal state standard once in fourteen years: and that the several county sealers shall compare their standard weights and measures with one of the state standards once, at least, in seven years: and the several town sealers shall compare their town standards with

APPENDIX. 199

the county standard once, at least, in three years: that, before either of the sealers of weights and measures, who shall be appointed by virtue of this act, shall enter on the duties of his office, he shall take and subscribe an oath or affirmation before one of the justices of the supreme court, or one of the judges of the court of common pleas, or justice of the peace of the county wherein such sealer is resident, well and truly, according to the best of his skill and ability, to perform the duties enjoined on him by this act: and that every assistant state sealer shall cause a certificate of the oath, by him taken, to be filed in the secretary's office: and every county sealer and town sealer shall, in like manner, cause such certificate as aforesaid to be filed in the clerk's office of their respective counties.

8. *And be it further enacted,* That each of the sealers of weights and measures, within this state, shall be entitled to receive for his services, in sealing and marking measures and beams which shall, from time to time, be brought to him for that purpose, twelve and an half cents, and for every weight, and every small liquid measure, three cents, over and above a reasonable compensation for making them conform to the standard established by this act.

9. *And be it further enacted,* That it shall be the duty of the clerks of the several counties to deliver to the respective county sealers of weights and measures heretofore appointed, or hereafter to be appointed, the standard brass yard measure which shall have been received by such clerks from the secretary of this state for the use of the said counties.

10. *And be it further enacted,* That no surveyor shall give evidence in any cause pending in any of the courts of this state, or before arbitrators, respecting the survey or measurement of lands, unless such surveyor shall make oath, if required, that the chain or measure used by him in surveying or measuring such lands was conformable to the standard measure of this state.

11. *And be it further enacted,* That, whenever either of the assistant state sealers of weights and measures shall resign or remove from the cities of New York or Albany, or the county of Oneida, or whenever any of the county or town sealers shall resign or remove from the counties or towns in which they were respectively appointed, it shall, in that case, be the duty of the person, so resigning or removing, to deliver to his successor in office all the standard beams, weights, and measures, in his possession; and, in case of the death of any sealer of weights and measures, it shall, in like manner, be the duty of his executors or administrators to deliver to the successor, to be appointed, all the said standard beams, weights, and measures, in the possession of their testator or intestate, at the time of his death; and, in case of neglect or refusal to deliver such standard entire and complete, the successor in office may maintain an action on the case against the person so removing or resigning, or against such executors or administrators, and recover double the value of such standard, or such parts thereof as have not been delivered to the said successor in office, with costs of suit; and, in every such action, if

judgment shall be rendered for the plaintiff, he shall recover double costs; one moiety of which may be retained by the person so recovering, and the other moiety shall be by him applied to the purchase of such standard beams, weights, and measures, as may not be delivered over as aforesaid.

12. *And be it further enacted,* That if any person or persons shall, after one year from the passing of this act, use any weights, measures, or beams, in weighing or measuring, which shall not be conformable to the standard of this state, whereby any purchaser of any commodity, or article of trade or traffic, shall be injured or defrauded, it shall be lawful for the person so injured or defrauded to maintain an action on the case against the offender in any court having cognizance thereof; and if judgment shall be rendered for the plaintiff, he shall recover treble damages against the defendant, with costs of suit.

E 7.

NEW JERSEY.

Elizabeth Town,
September 20, 1817.

Sir: I have the honor to acknowledge the receipt of your letter of the 29th of July last, accompanying a copy of a resolution of the Senate, relative to weights and measures. There is not in the state of New Jersey any legislative act establishing or regulating weights or measures, nor can I learn that the subject has, at any time, engaged the attention of the state legislature; but, by custom, the English standard of weights and measures has been adopted, and is in use throughout this state, and I know of no information that I can furnish, for enabling the Department of State to fulfil the views of the Senate.

I have the honor to be,
With very great respect,
Your most obedient servant,
ISAAC H. WILLIAMSON.

The Hon. Richard Rush,
Acting Secretary of State.

E 8—a.

PENNSYLVANIA.

Harrisburg, *August 26th, 1817.*

Sir: Your letter to the governor, accompanied by a copy of a resolution of the Senate of the United States, relative to weights and measures, has been duly received some time since. In this office there are no materials calculated to afford any information on that important subject. There does not appear to have been any legislative act relative to it since A. D. 1700, which directs the standards to be regulated " according to the king's standards for the exchequer," (see Smith's edition of the laws, page 19.) I addressed a note to Mr. Meer, keeper of weights and measures in the city: I herewith enclose his answer, together with a report of a committee of Senate in 1814, and Dr. Bollman's letter; none of which, probably, will suggest any new ideas to you on the subject.

Very respectfully, sir, your obedient servant,

N. B. BOILEAU, *Secretary.*

RICHARD RUSH, Esq.
Acting Secretary of State.

E 8—*b.*

PHILADELPHIA, *August* 20, 1817.

SIR: Your letter of the 16th inst. requests me to give information respecting weights and measures, for the consideration of the Senate of the United States—a task I cheerfully undertake, as far as I am acquainted with the business, because they are in an incorrect and deranged condition throughout the Union, in consequence of the regulators not having a fixed and correct standard, to which they can resort to keep those they use in order. The variant and irregular weights and measures of the different states and towns proves a serious evil, both to the wholesale and retail dealer, and often produces difficulties in the trade between the several states, which calls for immediate redress. It being the peculiar province of the legislature of the United States to fix the standard of weights and measures, I am happy to find that the Senate have taken the subject under their serious consideration.

The standards in my keeping were, I believe, brought here by William Penn, more than one hundred years ago, and have been in use ever since; of course they cannot be very correct.

I would therefore propose that one simple standard weight be adopted and used for all purposes, and that its scantlings and parts be decimally divided, so as to suit the money of the United States, and that the unit be the English avoirdupois pound.

That, for measures of capacity, the wine gallon be the unit, and be used for all purposes where measures of capacity are necessary; and that for the measure of extension the English foot be adopted as the unit, and be decimally divided.

I believe that the British standard weights are made of agate, so that they may not be corroded by oxidating principles of the air, as most metals are. But I believe that platina would be fitter for the purpose, being easily formed, and less liable than any other metal to be oxidated. I consider a cylindric form most suitable, both for weights and measures of capacity. No doubt the legislature will see the propriety of furnishing the capital or principal town of every state with a complete set of standards, so that the regulators may have the necessary standards made of coarser metal for their immediate use, as well as a resort to the means of keeping them in order.

In a trading community it is equally necessary to have correct scale beams as to have just weights. Frauds are daily occurring in consequence of not having a law for the inspection and regulation of them, similar to that for the regulation of weights. No person should be permitted to sell beams before they have been inspected and sealed by the proper officer.

I wish to refer you to Dr. Bollman's paper on the subject, in appendix to the journals of the Senate of March 18, 1814, No. 3; likewise to the report of March 3d, 1808, journals of the Senate; in which will be found valuable information.

I am, sir, with great respect, yours, &c.

JOHN MEER.

N. B. Boileau, Esq.

P. S. I wish to refer you, likewise, to a small treatise on moneys, coins, weights, and measures, proposed for the United States of America; wrote by Thomas Jefferson; printed by Daniel Humphries in 1789, Philadelphia.

I enclose Mr. Dorsey's report, as above noticed.

E 9—a.

DELAWARE.

Office of the Secretary of State,

Dover, Delaware, 7th Nov. 1818.

Sir: By direction of the Governor, I have the honor to reply to your letter requesting "such information of the acts of the state of Delaware, in relation to weights and measures, as he might think proper to communicate."

There is in our statute books but one act in relation to weights and measures—that is to be found in the first volume of our laws, page 57. By this act it is directed that standards of brass for weights and measures, according to the queen's standards for the exchequer, shall be obtained in each county within two years after the making of the law, and that these standards shall remain with such officer in each county as shall be, from time to time, appointed by the county court in each county; and that all weights and measures shall be made just, and marked by the keeper of these standards, &c. &c.

This law is not *now* observed in any part of the state, and I am unable to say whether it ever went into operation in the counties of Kent and Sussex; the probability is, that it never did, as no evidence can now be had that it ever was carried into effect in either of these counties. In the county of New Castle the act was carried into execution: the standard weights and measures which it prescribed were obtained for that county, and persons were appointed at different times to be keepers of those standards: but, for a great many years, the law has ceased to be observed in that county; nor is it known whether the standards that were procured for that county are now in existence.

The Philadelphia weights and measures are generally used, I believe, in this state, but whether they are conformable to the standards designated by the act of assembly aforesaid, or not, I cannot say.

No decision of any of our courts, sanctioning any particular weights or measures, has, to my knowledge, ever been made.

I herewith transmit you a communication which I lately received from James Booth, Esq. the chief justice of our courts of common pleas and quarter sessions, a gentleman whose age, experience, and different public stations, have afforded many opportunities for obtaining information as to the subject of weights and measures as used and regulated in this state.

I have the honor to be, sir,
With great respect,
Your most obedient servant,
H. M. RIDGELY.

The Hon. JOHN Q. ADAMS,.
Washington City.

E 9—*b.*

An Act for regulating Weights and Measures.

SEC. 1. *Be it enacted by the Honorable John Evans, Esq. with her Majesty's royal approbation, Lieutenaut Governor of the counties of New Castle, Kent, and Sussex, upon Delaware, and Province of Pennsylvania, by and with the advice and consent of the freemen of the said counties, in General Assembly met, and by the authority of the same,* That in each county of this her majesty's government there shall be had and obtained, within two years after the making of this law, at the charge of each county, to be paid out of the county levies, standards of brass for weights and measures, acccording to the queen's standards for the exchequer; which standards shall remain with such officer in the counties aforesaid as shall be, from time to time, appointed by the county court, in each respective county of this government: and every weight, according to its standard, and every measure, as bushel, half bushel, pecks, gallons, pottles, quarts, and pints, shall be made just weights and measures, and marked by him that shall keep the standards; and that no person, within this government, shall presume to buy or sell by any weights or measures not sealed or marked in form aforesaid, and made just according to the standards aforesaid by the officer in whose possession the standards remain, on penalty of forfeiting five shillings to the prosecutor, being convicted by one justice of the peace of the unjustness of his weights and measures; and that, once a year, at least, the said officer, with the grand jury, or the major part of them, and for want of a grand jury, with such as shall be appointed and allowed by the respective county courts aforesaid, for assistants, shall try the weights and measures in the counties aforesaid; and those weights and measures which are defective shall be seized by the said officer and assistants; which said officer for his fees, for his marking each bushel, half bushel, and peck, just measure, and marking the same that is large enough, when brought to his hands, shall have ten pence, and for every less measure, three pence; for every yard, three pence; for every hundred and half hundred weight, being made just and marked, three pence; for every less weight, one penny; and if the weights and measures be made just, before they be brought to him, then to have but half the fees aforesaid for marking the same. And if the said officer shall refuse to do any thing that is enjoined by this law, for the fees appointed, and be duly convicted thereof, shall forfeit five pounds, to the use of the governor for the time being. That a true measure or standard be taken from the brass half bushel in the town of Philadelphia, and bushel and a peck proportionable; and all less measures and weights coming from England, being duly sealed in London, or other measures agreeable therewith, shall be accounted and allowed to be good by the aforesaid officers, until the said standards shall be had and obtained.

SEC. 2. *And be it further enacted by the authority aforesaid,* That no person shall sell beer or ale by retail, but by beer measure, according to the standard of England.

E 9—c.

Mr. James Booth to H. M. Ridgely.

NEW CASTLE, *October* 24, 1818.

SIR: The letter, on the subject of weights and measures, which you were pleased to address to me, should have been earlier answered had not my engagements demanded my presence and attention to objects out of town.

The act of assembly passed in the reign of queen Anne, (vol. 1, 57,) to which you refer, was, I believe, carried into execution in the county of New Castle. The standard weights and measures which it prescribed, were obtained for this county; and I am enabled to state, from recollection, that persons were appointed, at two different times, to be keepers of those standards; other appointments were probably made, of which I have no recollection, and which cannot now be ascertained from the public records, many of which were lost during the revolutionary war. On inspecting those remaining in the office of the clerk of the peace, I found one appointment of keeper of the standards made in the year 1760. These keepers, with a part of the grand jury, traversed the county to examine and to rectify the weights and measures used by *sellers and buyers;* but this was not done once a year, agreeably to the act. I think I can recollect this duty to have been performed but twice in different years; and it is so long since, that I cannot pretend to point out the year when it was last performed; nor can I tell whether these standards are still in existence; or if they are, in whose possession they remain. I am also unable to state, whether the act of assembly went into operation, or not, in the counties of Kent and Sussex, or in either of them.

Whether the Philadelphia weights and measures differ from the exchequer standards, I cannot tell; but, I believe, the Philadelphia measures of capacity, particularly, are generally used in this state. I think it is so in the county of New Castle; and I remember that, in a controversy in Sussex about corn, it appeared from the testimony, that the Philadelphia sealed bushel was deemed by the parties to be the proper measure to ascertain the quantity. But I know of no decision of any of our courts, sanctioning any particular weights or measures.

I regret that I can give you no definite information on all the subjects of your inquiry, nor do I know any source from which it can be drawn.

With every sentiment of regard,
    I, am sir, &c.

JAS. BOOTH.

H. M. Ridgely, Esq.

E 10—*a*.

MARYLAND.

Council Chamber, Annapolis, August 9, 1817.

Sir: In the absence of his excellency the governor from the state of Maryland, I do myself the honor to acknowledge the receipt of your letter of the 29th ultimo, at this department, and to inform you that the English standard of weights and measures has been adopted by this state. His excellency, however, will do himself the honor, immediately on his return, of replying to your letter.

I have the honor to be,
With high respect,
Your most obedient servant,
NINIAN PINKNEY,
Clerk of the Council.

The honorable RICHARD RUSH,
Acting Secretary of State.

E 10—b.

COUNCIL CHAMBER, ANNAPOLIS,
December 1, 1819.

SIR: In reply to your letter of the 1st of November, enclosing a copy of a resolution of the Senate of the United States, passed the 3d of March, 1817, on the subject of weights and measures, I have the honor to inform you that the only legal regulation upon those subjects in the state of Maryland is comprised in several acts of Assembly, of which copies are herewith transmitted to you. It may be proper to add that, by the act incorporating the city of Baltimore, passed at November session, 1796, chapter 68, the powers before vested by law in the standard keeper of Baltimore county, were transferred, within the limits of their jurisdiction, to that corporation, whose proceedings, from its being the only large commercial town we have, in effect now regulate the weights and measures throughout the whole state. It is believed that the English standard of measures has not, in practice, been strictly adhered to; as it is recollected that some years ago the half bushel which was used for measuring grain at Elkton, was larger than the one in use at Baltimore; and that the same measure in Baltimore was somewhat larger than the old standard kept in the different counties. It is believed, also, that a similar difference, in a small degree, exists between the fifty-six and other weights, regulated in Baltimore, and those adjusted by the standard keepers in the counties. The result has been that the Baltimore standard, both of weights and measures, now governs all the dealings and business of the state.

I am, sir, with much respect,
Your obedient servant,
C. GOLDSBOROUGH.

Hon. JOHN QUINCY ADAMS,
Secretary of State.

APPENDIX.

E 10—c.

An act relating to the standard of English weights and measures. Passed April, 1715.

Whereas the standards of English weights and measures are very much impaired in several of the counties of this province, and in some wholly lost, or unfit for use:

Be it enacted by the king's most excellent majesty, by and with the advice and consent of his majesty's governor, council, and assembly, of this province, and by the authority of the same, That the justices of the several county courts shall, by all convenient speed, at the charge of their respective counties, cause the standards they already have to be made complete, and purchase new standards where they have none; and, for the better preservation of them for the future, that they take good and sufficient security, in his majesty's name, to the use of the county where taken, from the persons that shall be entrusted by them to keep each standard, in the penal sum of fifty pounds sterling "for the safe-keeping of such standard, and for the due execution of the office of standard-keeper, and for the delivering the same up in the like good order they receive the same, when they shall be legally discharged from such trust," under the penalty of five hundred pounds of tobacco for each justice of that county court that shall omit to do what is required of them by this act, the one half to his majesty, his heirs and successors, for the support of government, the other half to the informer, or to him or them that shall sue for the same, to be recovered in the provincial court of this province, against such justices, jointly or severally, by action of debt, bill, plaint, or information, wherein no essoin, protection, or wager of law to be allowed.

3. *And be it enacted by the authority aforesaid, by and with the advice and consent aforesaid,* That all persons, whether inhabitants or foreigners, shall repair and bring their steelyards with which they weigh and receive their tobacco, to the standard, yearly and every year, to be tried, stamped, and numbered, for which they are to pay the person keeping the standard one shilling for every time such steelyards shall be tried and stamped as aforesaid; and every person or persons shall have their bushel, half bushel, peck, gallon, pottle, quart, and pint, if they make use of the same, or any of them, in buying or selling, duly tried and stamped at the standard aforesaid, except such of the measures aforesaid as come out of England, and are there stamped; for which trying and stamping they shall pay six pence a piece.

4. And whosoever shall presume to sell by any dry measures, without first having the said measures tried and stamped at the standard, shall forfeit the sum of five hundred pounds of tobacco.

5. And whosoever, likewise, shall presume to weigh and receive tobacco by steelyards which have not within one year past from such weighing and receiving, been tried and stamped at the standard,

shall forfeit one thousand pounds of tobacco, the one half of which aforementioned forfeitures to be paid to his majesty, his heirs and successors, towards the defraying the charge of the county where the offender shall dwell or reside, and the other half to the informer or informers, to be recovered in any county court of this province, by bill, plaint, or information, wherein no essoin, protection, or wager of law to be allowed.

6. And, if any person or persons shall refuse to pay any tobacco by such steelyards, tried and stamped as aforesaid, and shall thereby compel the owner to have them tried over again within the year, if the steelyards are true, such person, so refusing or compelling as aforesaid, shall pay for the new stamping, but if not, the owners of the steelyards to pay for the same.

E 10—d.

A supplementary act to the act, entitled " an act relating to the standard of English weights and measures." Passed 20th December, 1765.

Whereas, by the act entitled an act relating to the standard of English weights and measures, no penalty is imposed upon persons buying by any dry measure or measures, without first having the said measure or measures tried and stamped at the standard, as there is upon the seller: And whereas it is represented to this General Assembly that many buyers of grain, flaxseed, and other commodities, when the people have carried them a great distance to market, refuse to buy them unless by measure or measures of their own, which have been found, upon trial, to be larger than the standard aforesaid :

2. *Be it therefore enacted by the right honorable the lord proprietary, by and with the advice and consent of his lordship's governor, and the upper and lower houses of assembly, and the authority of the same,* That, if any person or persons shall hereafter buy, by any dry measure or measures, being his, her, or their property, or found or provided by him, her, or them, contrary to the true intent and meaning of the above recited act, he, she, or they, shall forfeit and pay the sum of five pounds current money for every offence, one half thereof to the informer, or him, her, or them, that shall sue for the same, and the other half to be paid into the hands of the treasurer of the respective shore where such forfeiture shall happen, and applied as the General Assembly for the time being shall direct and appoint, to be recovered in any court of record within this province, by action on the case, action of debt, bill of indictment or information, wherein no essoin, protection, or wager of law, or more than one imparlance, shall be allowed.

3. *Provided always,* That such action shall be commenced within one year from the time of the said offence being committed, and not afterwards.

APPENDIX. 209

E 11—*a*.

VIRGINIA.

The Governor of Virginia to the Secretary of State.

RICHMOND, VIRGINIA, *Council Chamber, Aug.* 15, 1818.

SIR: In compliance with the request contained in your letter of the 5th instant, I have the honor herewith of transmitting you the only laws on the statute book of this state on the subject of weights and measures.

I presume this is all the information which you need, or would wish to have communicated.

I have the honor to be, with the greatest respect,
Your obedient servant,
JAMES P. PRESTON.

The Hon. JOHN QUINCY ADAMS,
Secretary of State.

E 11—*b*.

An act for more effectually obliging persons to buy and sell by weights and measures according to the English standard. Passed August, 1734.

1. Forasmuch as the buying and selling by false weights and measures is of late much practised in this colony, to the great injury of the people:

2. *Be it enacted by the lieutenant governor, council, and burgesses, of this present assembly, and by the authority of the same,* That from henceforth there shall be but one weight, one measure, one yard, and one ell, according to the standard of the exchequer in England : and whosoever shall sell or lay by, or keep, any other weight, measure, yard, or ell, whereby any corn, grain, salt, or other thing, is bought or sold, after the tenth day of June, one thousand seven hundred and thirty-six, shall forfeit, for every offence, twenty shillings, being thereof lawfully convicted by the oath of one sufficient witness, before any justice of the peace of the county where the offence shall be committed; to be levied, by distress and sale of the goods of the offender, for the use of the poor of the parish; rendering the overplus to the party so offending: and in default of such distress, such justice of the peace shall commit the said party to the common *gaol* or prison, there to remain without bail or mainprize until he shall pay such forfeitures as aforesaid.

3. And to the end, all people may be more easily provided with such weights and measures, *Be it further enacted,* That the justices of the peace of every county, where they have not already provided the

same, shall, within eighteen months after the end of this session of assembly, provide, at the charge of their respective counties, brass weights, of half hundreds, quarters, half quarters, seven pounds, four pounds, two pounds, and one pound weight, according to the said standard: and measures of bushel, half bushel, peck, and half peck, dry measure, according to that standard: and gallon, pottle, quart, and pint, of wine measure, according to the said standard; with proper scales for the weights, upon pain of forfeiting, by every justice sworn into the commission of the peace, five shillings for every month such weights and measures shall be wanting, to be recovered, by action of debt, or information in any court of record in this colony; one half whereof shall go to the king, his heirs, and successors, for supporting the contingent charges of this government, and the other moiety to the informer.

4. And the said weights and measures, so to be provided, shall be kept, from time to time, by such person as shall be appointed by the county courts respectively, to which all persons may resort for trying their weights and measures; and when they are tried, and found to agree with the standard, the same shall be sealed by the person keeping such standard, with a seal, to be likewise provided by the justices aforesaid. And that the fees to the persons entrusted with the keeping such standard weights and measures, be, for the trying every *stillyard* and certificate thereof, one shilling; for the trying any weights or measures and sealing the same, four pence for every such weight or measure sealed, to be paid by the person for whom the service shall be done; any former law, custom, or usage, to the contrary hereof, in any wise, notwithstanding.

5. *Provided always,* That this act, or any thing herein contained, shall not be construed to prohibit any person or persons whatsoever from buying and selling by steelyards, which shall be tried by and agree with the standard aforesaid, where the buyer and seller, payer and receiver, shall both consent thereto: any thing in this act contained to the contrary hereof notwithstanding.

A copy.
Test,
ANTHY. WHITAKER, *Cop'g C. C.*

E 11—c.

An act concerning weights and measures. Passed 26th December, 1792.

1. Whereas the general assembly of Virginia, at their session in the year 1734, did pass an act, entitled "An act for more effectually obliging persons to buy and sell by weights and measures according to the English standard:"

2. *Be it therefore enacted by the general assembly,* That the said act, shall continue and remain in force until the Congress of the United States shall have made provision on that subject.

APPENDIX.

3. *Provided always,* That all fines, forfeitures, and penalties, in the said ac mentioned, shall be and enure one moiety to the commonwealth, and the other to the use of the informer.

4. This act shall commence, and be in force, after the passing thereof.

A copy.
Test,
ANTHY. WHITAKER, *Cop'g C. C.*

E 12—a.

NORTH CAROLINA.

Governor Miller to Mr. Rush, acting Secretary of State.

EXECUTIVE OFFICE, N. C.
Raleigh, August 19, 1817.

SIR: In compliance with the request contained in your letter of the 29th July, I enclose you a certified copy of an act of the general assembly of this state regulating weights and measures.

With much respect,
Your obedient servant,
WILLIAM MILLER.

RICHARD RUSH, Esq.
Acting Secretary of State.

E 12—b.

An act for regulating weights and measures.

1. Whereas many notorious frauds and deceits are daily committed by false weights and measures: for prevention whereof,

2. *We pray that it may be enacted, and be it enacted by his excellency Gabriel Johnson, Esq. Governor, by and with the advice and consent of his majesty's Council and General Assembly of this province, and it is hereby enacted by the authority of the same,* That no inhabitant or trader shall buy or sell, or otherwise make use of in trading, any other weights or measures than are made and used according to the standard in his majesty's exchequer, and the statutes of England in that case provided.

3. *And, for the discovery of abuses, be it further enacted by the authority aforesaid,* That the justices of each and every county within this

government shall, within two years next after the ratification of this act, at the charge of each county, respectively, provide sealed weights of half hundred, quarters of hundreds, seven pounds, four pounds, two pounds, one pound, and half pound; and measures of ell and yard, of brass or copper, and measures of half bushel, peck, and gallon of dry measure ; and a gallon, pottle, quart, and pint, of wine measure; for the payment of which charge the said justices are hereby empowered to levy a tax on their respective counties, to be kept by such person, and in such place, as the justices of each respective county shall appoint, such person first giving sufficient security to the said justices in the sum of fifty pounds proclamation money ; and the said justices shall also find and provide for the said person, a stamp for brass, tin, iron, lead, or pewter weights or measures, and also a brand for wooden measures, of the letters N C, upon pain of forfeiting and paying the sum of ten pounds proclamation money, to be recovered from the said justices by action of debt, bill, plaint, or information, in the general court of this province, and applied to the use of our sovereign lord the king, for and towards the support of this government and the contingent charges thereof.

4. *And be it further enacted by the authority aforesaid,* That any person whatever using weights and measures, shall bring all their measures and weights to the keeper of the standard of the county where such person shall reside or trade, to be there tried by the standard and sealed or stamped : and if any person, or persons, shall buy, sell, or barter, by any weight or measure which shall not be tried by the standard, and sealed or stamped as aforesaid, he, she, or they, so offending, shall, for every such offence, forfeit and pay the sum of ten pounds proclamation money, one half to the use of the county where such offence shall be committed, and the other half to the use of the party who shall sue for the same, to be recovered in any court of record in this government, wherein no essoign, protection, privilege, injunction, or wager of law, shall be allowed.

5. And whereas steelyards by use are subject to alterations :

Be it further enacted by the authority aforesaid, That all and every person, who shall use, buy, or sell, by steelyards, shall, once every year, try the same with the standard, and take a certificate from the keeper of the standard for the county wherein such person shall reside, upon pain of twenty shillings proclamation money, to be recovered and applied as aforesaid.

6. Repealed, vol. 2, 48.

7. *And be it further enacted by the authority aforesaid,* That the standard keeper of each and every county shall, at the next court to be held for the county in which he shall reside, take the following oath, viz: " You shall swear that you will not stamp, steal, or give
" any certificate for any steelyards, weights, or measures, but such
" as shall, as near as possible, agree with the standard in your keep-
" ing, and that you will, in all respects, truly and faithfully dis-
" charge and execute the power and trust by this act reposed in you.
" to the best of your ability and capacity. So help you God."

8. *And be it further enacted by the authority aforesaid*, That the standard keeper of each and every county in this government, is hereby empowered and required, with the assistance of a constable (who is hereby commanded upon notice to attend him, upon information made to him of any person, or persons, keeping, or having in his or their house, or custody, any steelyards, weights, or measures, which shall have been altered, lessened, or shortened, since they were tried and sealed by the standard, or shall be suspected of buying, selling, or bartering, by such false weights and measures) to search the houses or other suspected places of such offender, for any such weights or measures so falsified ; and if, upon search, any such false weights or measures shall be found, he shall charge a constable with the owner of them, or the person using them, who shall forthwith convey him, her, or them, before any justice of the peace, who is hereby directed to bind him, her, or them, over to the next court to be held for the county where the offence shall be committed ; and the said offence shall be laid before the grand jury by the king's attorney general, or his deputy, and for want of them, by any person the county court shall think fit to appoint, and shall be cognizable by the said grand jury either by indictment or presentment ; and if upon trial by a petit jury such offender or offenders shall be found guilty, the county court shall fine each and every person, so convicted, in a sum not exceeding twenty-five pounds proclamation money, one-third part thereof to the informer, one-third part to the standard keeper, and one-third part thereof to be paid to the justices of the county, to be applied to the use of the county where the offence shall be committed ; and shall commit the offender to jail until the same shall be paid: and further, if it appear to the county court, by the verdict of the petit jury, that the offender altered, lessened, or shortened, his or her steelyards, weights, or measures, or caused the same to be done, or used such steelyards, weights, or measures, knowingly, after they were so altered, lessened, or shortened, with an intent to defraud any person, in such case, the court shall, besides and notwithstanding the said fine, sentence such offender to stand publicly during the sitting of the court two hours in the pillory, with his offence written over his or her head; any law, custom, or usage, to the contrary notwithstanding.

9. *And be it further enacted by the authority aforesaid*, That the naval officer of each and every port within this government shall affix up, in a public part of his office, and there constantly keep affixed, an advertisement of this act, that traders coming into this government may have notice thereof, upon pain of forfeiting five shillings proclamation money, for every twenty-four hours the same shall be neglected, to be recovered by a warrant from any justice of the peace of the county where the offence shall be committed, by any person who shall sue for the same, and applied, one-half to the informer, and the other half to the use of the said county.

10. *And be it further enacted by the authority aforesaid*, That the justices of every county, respectively, shall have power to take and

receive into their custody all such weights and measures as have been already provided by their respective county or parish, and shall also demand and receive from all and every person, or persons, whatsoever, all such sums of money as have been already raised to purchase such weights and measures, and dispose of and apply the same according to the directions of this act.

11. *And be it further enacted by the authority aforesaid,* That all and every other act, and acts, and every clause and article thereof, heretofore made, so far as relate to weights and measures, or any other matter or thing within the purview of this act, is, and are hereby, repealed and made void, to all intents and purposes, as if the same had never been made.

A true copy.
Given 18th August, 1817.

WM. HILL, *Secretary.*

E 13—a.

SOUTH CAROLINA.

The Governor of South Carolina to the Secretary of State.

EXECUTIVE OFFICE, SOUTH CAROLINA,
10*th September,* 1818.

SIR: Yours of the 5th ultimo, covering a resolution of the Senate of the United States, requesting information upon the laws of this state regulating weights and measures within the same, has been received. I had the honor of receiving a former like communication from you, and directed the secretary of this state to give you every information upon the subject. This, I presume, has never reached you. I now enclose you copies of two sections of acts, which, I believe, are every thing contained in our statute book relating to weights and measures.

I have the honor to be,
With great respect, &c.

ANDREW PICKENS.

Honorable SECRETARY OF STATE
   for the United States.

APPENDIX.

E 13—b.

Act of Assembly, passed 12th April, 1768.

"SEC. 5. The public treasurer shall, immediately after the passing of this act, procure, or cause to be made, of brass or other proper metal, one weight of fifty pounds, one of twenty-five pounds, one of fourteen pounds, two of six pounds, two of four pounds, two of two pounds, and two of one pound, avoirdupois weight, according to the standard of London; and, also, of cedar wood, neatly shaped and handled with iron, one bushel, one half bushel, one peck, one half peck measures, according to the standard of London; which weights shall each, respectively, be stamped or marked in figures denominating the weight thereof, and shall be kept by the said public treasurer: and the said weights and measures shall be deemed and taken to be the standard weights and measures by which all the weights and measures in this province shall be regulated."

Act passed 17th March, 1785.

"SEC. 63. The several justices of the county courts in this state, as soon as the same shall take place in the respective counties, shall have full power and authority to regulate weights and measures within each of their respective jurisdictions, and shall enforce the observance thereof in such manner and form, and under such penalties, as are already prescribed by law for regulating weights and measures."

E 14—a.

GEORGIA.

The Governor of Georgia to the Secretary of State.

EXECUTIVE DEPARTMENT, GEORGIA,
Milledgeville, 5th December, 1819.

SIR: I have the honor to enclose a copy of an act of the general assembly of this state, and an extract from an ordinance passed by the corporation of the city of Augusta, showing the regulation and standard for weights and measures in this state. Some delay in obtaining from Augusta the extract, together with a multiplicity of business occasioned by the legislature now in session, has prevented me from complying with your request at an earlier period.

The documents mentioned in your letter of the 4th ultimo have been duly received at this department.

I have the honor to be,
Very respectfully,
Your most obedient servant,
JOHN CLARK.

Honorable JOHN QUINCY ADAMS,
Secretary of State.

E 14—b.

An Act to regulate weights and measures in this state.

SEC. 1. *Be it enacted by the Senate and House of Representatives of the state of Georgia, in general assembly met, and by the authority of the same,* That the standard of weights and measures established by the corporations of the cities of Savannah and Augusta, and now in use within the said cities, shall be, and the same are hereby declared to be, the fixed standard of weights and measures within this state; and all persons buying or selling shall buy and sell by that standard, until the Congress of the United States shall have made provision on that subject.

SEC. 2. *And be it further enacted by the authority aforesaid,* That it shall be the duty of the justices of the inferior courts, or a majority of them, in their respective counties, by their clerk, or some other person especially authorized by them for that purpose, to obtain from the said corporations, or one of them, to be paid out of the county funds, the standard of weights and measures as fixed by them, within six months from the passing of this act; and that the said justices, or a majority of them, shall, so soon as they obtain the standard of such weights and measures, give thirty days' notice thereof at the court-house and three other public places in the county; and if any person or persons whosoever shall sell, or attempt to sell, any article or thing by any other or less weight or measure than that so established, he, she, or they, so offending, shall forfeit and pay three times the value of the article so sold, or attempted to be sold, to be recovered before any justice of the peace, if it should not amount to more than thirty dollars, and if above that sum before any judge of the superior court, or the justices of the inferior court, by action of debt; one half whereof shall be for the use of the informer or person bringing the action, and the other for the use of the county in which such act or offence may happen.

SEC. 3. *And be it further enacted by the authority aforesaid,* That it shall be the duty of the justices of the inferior court, or a majority of them, of the respective counties of this state, to procure a marking instrument, seal, or stamp, for the purpose of marking, sealing, or stamping, all weights and measures within their several counties, which marking instrument, seal, or stamp, shall remain in the clerk's office of the inferior court, by him to be affixed to any weight or measure which he may find to correspond with or not less than the standards established by said corporations of Savannah and Augusta.

SEC. 4. *And be it further enacted,* That the said clerks of the inferior courts shall receive six and one-fourth cents for every weight and measure by them so marked, sealed, or stamped, to be paid by the person obtaining the same.

ABRAHAM JACKSON,
Speaker of the House of Representatives.
DAVID EMANUEL,
President of the Senate.

Assented to, December 10, 1803.
JOHN MILLEDGE, Governor.

E 14—c.

"*And be it further ordained,* That all weights for weighing any articles of produce or merchandise shall be of the avoirdupois standard weight; and all measures for liquor, whether of wine or ardent spirits, of the wine measure standard; and all measures for grain, salt, or other articles usually sold by the bushel, of the dry or Winchester measure standard: and all weights and measures used within this city shall be in conformity to the said public standard. It shall not be lawful for any person or persons to make use of any other than brass or iron weights, regulated as aforesaid, or weights of any other description than those of fifty, twenty-five, fourteen, seven, four, two, one, half, quarter pound, two ounce, one ounce, and downwards."

I certify the above to be true extracts from the original ordinance of the city council of Augusta.

Clerk's Office, 30th *November,* 1819.
JOSEPH CRANE, *Dep. C. C.*

E 15—a.

KENTUCKY.

The Governor of Kentucky to the Secretary of State.

FRANKFORT, *November* 26, 1817.

SIR: In compliance with your request, and the resolution of Congress on the subject of weights and measures, I have the honor to transmit to you a copy of the law of this state on that subject.

It may not be improper to remark, that weights and measures, agreeably to said law, have been procured as a general standard for this state.

I have the honor to be,
Your most ob't and humble servant,
GABRIEL SLAUGHTER.

The Hon. SECRETARY OF STATE,
Washington City.

E 15—*b.*

An Act concerning Weights and Measures, approved December 11, 1798.

Whereas the Congress of the United States are empowered, by the federal constitution, to fix the standard of weights and measures, and they have not hitherto passed any law for the aforesaid purpose: whereby an act passed by the general assembly of Virginia, in the year 1734, entitled "An act for more effectually obliging persons to buy and sell by weights and measures according to the English standard," still remains in force in this commonwealth:

SEC. 1. *Therefore, be it enacted by the General Assembly,* That the governor be, and he is hereby, authorized and directed to procure one set of the weights and measures in the said act specified, with proper scales for the weights, together with measures of the length of one foot and of one yard; and the bushel, dry measure, shall contain two thousand one hundred and fifty and two-thirds solid inches; and the gallon of wine measure shall contain two hundred and thirty-one inches; and the said weights, measures, and scales, shall be deposited in the custody of the secretary of state, to serve as a general standard for weights and measures within this commonwealth.

SEC. 2. *And be it further enacted,* That, when the aforesaid weights, measures, and scales, shall be procured as aforesaid, the governor shall cause to be made for each county, within this state, one set of weights and measures, and the last mentioned weights and measures shall be compared by the secretary of state with the aforesaid general standard, and, if found to agree therewith, shall be forthwith transmitted by him, together with scales proper for the weights to be procured as aforesaid, to the clerks of the several county courts in this state.

SEC. 3. *And be it further enacted,* That the said weights, measures, and scales, shall be kept by such person, in each county, as the court of the said county shall appoint; and, immediately after such appointment, the clerk shall make known the same by advertisement, to be fixed up at the door of the court house. And all persons, desirous of trying their weights and measures, may resort to the aforesaid county standards for that purpose; and the person appointed to keep the said standards shall, if he find them true, seal them with a seal, to be provided by the county court, at the expense of the county: and the persons appointed in the several counties to keep the said county standards shall be entitled, for trying every steelyard, and certificate thereof, to twenty-five cents; for trying any weights or measures, and sealing the same, twelve and one half cents, for each weight or measure sealed, to be paid by the person for whom such service shall be done.

SEC. 4. *And be it it further enacted,* That, three months after the appointment of a person to keep the said county standards shall have been made known as aforesaid, every person who shall knowingly buy, or who shall sell, any commodity whatever by weight or measure that shall not correspond with the said county standards, or shall keep any such for the purpose of buying or selling with them, shall, for every such offence, forfeit and pay four dollars, to go towards lessening the county levy, and to be recovered before any justice of the peace for the county in which such offence shall be committed.

SEC. 5. *And be it further enacted,* That the auditor of public accounts shall issue his warrant, or warrants, on the treasurer, to such amount as the governor shall certify to be due, and to such person or persons as the same shall be owing, for furnishing and transporting the aforesaid weights, measures, and scales.

This act shall be in force from and after the passage thereof.

E 16.

TENNESSEE.

The Governor of Tennessee to the Secretary of State.

MURFREESBORO', *November* 24, 1819.

SIR: Your communication, relative to a standard of weights and measures, has been received, and as no fixed standard exists as to weights, I laid your request before the legislature, who are now in session—indeed I pursued the same course two years past, when a similar request was made by Mr. Rush, and am sorry to say the subject was not attended to, which, I trust, will not be the case in the present instance. At all events, I will do myself the honor to advise you of the result.

I am, with great respect,
Your obedient servant,
JOS. M'MINN.

JOHN QUINCY ADAMS, Esq.
Secretary of State of the United States.

APPENDIX.

E 17—a.

OHIO.

Governor Worthington to Mr. Rush, acting Secretary of State.

CHILLICOTHE, 18*th August*, 1817.

SIR: In compliance with the request made in your letter of the 29th ult. I have directed the secretary of state of Ohio to forward a certified copy of any law of the state for the regulation of weights and measures.

I have the honor to be, very respectfully,

T. WORTHINGTON.

RICHARD RUSH,
Acting Secretary of State of the United States.

E 17—b.

COLUMBUS, (Ohio,) 13*th September*, 1817.

SIR: Pursuant to a request of the governor of our state, I have the honor to transmit an extract from an act of this state, relating to measures. I concluded it was not necessary to transcribe the whole act, as the remainder thereof only relates to the appointment of a person to keep said standard, and to try and prove others thereby. As to weights, we have no legislative provision on the subject, which is much to be regretted.

I have the honor to be, with great respect,
Your most obedient, humble servant,

JOHN M'LENE.

The Hon. SECRETARY OF STATE,
for the United States.

E 17—c.

Extract from an act of the General Assembly of the state of Ohio, entitled "An act for regulating measures," passed the 22d day of January, 1811, and now in force.

"SEC. 1. *Be it enacted by the General Assembly of the state of Ohio,* That the county commissioners of each county in this state are hereby required and directed to cause to be made for each county one half bushel measure, which shall contain one thousand seventy-five and two tenths solid inches, which shall be kept in the county seat, and shall be called the standard."

APPENDIX.

E 18—a.

LOUISIANA.

New Orleans, September 15*th*, 1817.

Sir: I have the honor to acknowledge the receipt of your letter of the 29th of July last, together with the resolve of the Senate of the United States, and to enclose you herein the true copy of an act passed in 1815, by which you will perceive that the legislature of this state have deemed most proper to adopt the same standards for weights and measures as established in the United States, and used by their revenue officers.

With sentiments of very great respect,
I remain, sir, &c.
JAS. VILLERE,
Governor of the state of Louisiana.

The Hon. Richard Rush.

E 18—b.

An act to establish an uniform standard of weights and measures within this state.

Whereas it is essential to the commerce of the state of Louisiana, that an uniform standard of weights and measures be established by law; therefore,

Sec. 1. *Be it enacted by the Senate and House of Representatives of the state of Louisiana in general assembly convened,* That the governor, at the expense of the state, shall procure one complete set of copper weights, to correspond with weights of their like denomination, used by the revenue officers of the United States in their offices, together with scales for said weights, and a stamp or seal, with such device as the governor may choose; as also one complete set of measures, calculated for dry, liquid, and long measure, of the same capacity and length as those of their like denomination used by revenue officers as aforesaid; which set of weights and measures, together with the scales and stamp, to be deposited in the custody of the secretary of state, to serve as a general standard of weights and measures within this state.

Sec. 2. *And be it further enacted,* That the administration and distribution of weights and measures within the limits of the city of New Orleans, shall be confided to an inspector of weights and measures, to be nominated by the governor, with the advice and consent of the senate; and it shall be the duty of the said inspector to see that no other weights but those established by this act be made use of within the limits of the said city, and in case of negligence, or breach on the part of the said inspector, he shall be condemned to pay a fine not exceeding two hundred dollars, nor less than one hundred; the mayor and city council of New Orleans shall be authorized to pass any regulations, or such ordinances relative to the police of the said weights and measures, and to insure, within the limits of the said city, the execution of the present act; provided, however, that said ordinances or regulations do not contravene the provisions of this act, and that, should the office of inspector become vacant during the recess of the legislature, it shall be the duty of the governor to fill such vacancy.

SEC. 3. *And be it further enacted,* That every parish judge shall procure, at the expense of his parish, a set of weights and measures, and a stamp, conformable to those mentioned in the first section of this act, the same to be stamped, on their request, by the secretary of state, or his deputy;—the inspector of the city of Orleans shall procure the above set at the expense of the corporation of the said city.

SEC. 4. *And be it further enacted,* That the aforesaid weights, measures, and stamp, shall be deposited by the judge in the office of the clerk of the said parish on his accountable receipt, and it shall be the duty of the said clerk and inspector, when thereto required by any person, to stamp all measures whatsoever, if they find them true; and they shall be entitled to ask and demand for such service the following fees, and no more, viz: For every steelyard, with certificate thereof, twenty-five cents; for every measure that they shall try or stamp, six cents and a quarter, to be paid by the owners of the weights and measures by them stamped, over and above the price of the labor, and materials they may employ on such measures as require to be regulated by the standard; provided always, that the stamp shall be impressed and payment required for doing the same, only on such as have not yet been stamped, or such as having been once stamped, are found so defective as to require to be again regulated with the standard. Be it understood, however, that during the first year ensuing the day when this act shall begin to be in force, the said inspector shall only be entitled to half the fees established by the present section, and in case of resignation, removal, or death of the inspector, or clerks, the said weights, measures, steelyards, and stamps, shall be delivered to the person or persons named in their places.

SEC. 5. *And be it further enacted,* That if any clerk, inspector, or any person legally authorized to stamp weights and measures, shall knowingly and wilfully stamp weights and measures which do not correspond with the standard aforesaid, they shall, on conviction, be condemned to pay a fine of one hundred dollars for each offence, to be recovered on motion by the attorney general, or district attorney, before any court of competent jurisdiction, to the benefit of the parish in which the said offender may reside; and any person thus convicted shall besides be removed from office: the court, during the prosecution against the said clerks, shall be authorized to appoint a clerk pro tem.

SEC. 6. *And be it further enacted,* That one year after the governor shall have deposited such weights, measures, and steelyards, with the secretary of state, as required by this act, no person shall buy or sell any commodity whatsoever by weights or measures, which do not exactly correspond with the aforesaid standard, or are not stamped; nor shall they keep any such weights or measures for the purpose of buying or selling thereby, under penalty of a fine of fifty dollars for each offence, besides the forfeitures of the weights, measures, and steelyards, found to be false, and of a fine of ten dollars, when the measures, weights, and steelyards, shall be found just, though not stamped; said fines to be recovered before any tribunal of competent jurisdiction, one half to the benefit of the informer, and the other half to the parish in which the said offender may reside. All weights and measures seized shall be forfeited for the benefit of the stamper, who shall have discovered the fraud, and he shall not return them into circulation until he has made them conformable to the standard.

SEC. 7. *And be it further enacted,* That whoever shall make, or cause to be made use of, or utter false stamps or scales, shall, on conviction before the district court where this offence is committed, be subject to all the pains and penalties of forgery under the law of this state.

SEC. 8. *And be it further enacted,* That it is forbidden to any person to sell, or cause to be sold, measures and weights, unless they have been tried and stamped by persons appointed for that purpose, agreeably to the present act, under the penalties imposed by the sixth section of this act, against all persons who shall have used false weights.

SEC. 9. *And be it further enacted,* That the persons to be appointed agreeably to the present act to try and stamp the weights, measures, &c. shall not commit their functions to a substitute, without being liable to all fines prescribed by this act.

SEC. 10. *And be it further enacted,* That any provision in this act contained to the contrary notwithstanding, there shall be in this state a dry measure to be known under the name of barrel, which shall contain three and a quarter bushels, conformable to the American standard, and shall be divided in half and a quarter barrel.

SEC. 11. *And be it further enacted,* That it shall be the duty of the governor, so soon as he shall have procured the weights and measures agreeably to this act, to make it known throughout the state by proclamation.

MAGLOIRE GUICHARD,
Speaker of the House of Representatives.
FULWAR SKIPWITH,
President of the Senate.

Approved, Dec. 21st, 1814:
WILLIAM C. C. CLAIBORNE,
Governor of the state of Louisiana.

A true copy from the original in my office.
MAZUREAU, *Secretary of State.*

APPENDIX.

E 18—c.

[TRANSLATION.]

Statement of M. Bouchon, respecting weights and measures in Louisiana before its cession to the United States.

1st. The measures in use in Louisiana, before its cession to the United States, were the following:

Measures of Length.

The king's foot, consisting of 12 inches, the inch of 12 lines;
The fathom (toise) of 6 king's feet;
The perch of 3 fathoms, or 18 feet;
The linear arpent of 10 perches, or 30 fathoms;
The league of 84 arpens;
The ell of Paris of 3 feet, 7 inches, $10\frac{5}{6}$ lines.

Superficial Measures.

The square foot of 144 square inches;
The square fathom of 36 square feet;
The square perch of 9 square fathoms;
The square arpent of 100 square perches;
The square league of 7,056 square arpens.

Weights.

The scruple of 24 grains;
The gros of 3 scruples
The ounce of 8 gros;
The mark of 8 ounces,
The pound of 16 ounces;
The quintal of 100 pounds.

Measures of Capacity for Liquids.

The pint of Paris of 46.95 cubic inches;
The velte of 8 pints;
The muid of 288 pints.

Measures for Grain.

The litron of Paris of 40.986 cubic inches;
The bushel (boisseau) of 16 litrons;
The septier of 12 bushels;
The muid of 12 septiers.

Measures for Fire Wood.

The cord of 4 feet height, 8 feet length, and the billets of wood 4 feet long.

2d. These measures are all of French origin. The lineal arpent, the league of 84 arpens, and the square arpent, are the particular measures of Louisiana. This is certain.

3d. The measures of length, of superficies, and of firewood, are still in general use in the state of Louisiana. They begin to measure firewood by the American foot, because it is to the advantage of the seller. For the same reason, the use of the pound has been so easily adopted. I know of no other measure of capacity, except the American gallon. Several persons who speculate in town lots buy by the French foot and sometimes sell by the American foot. They sometimes yet use the French ell, but its use in commerce is insensibly losing ground.

4th. The surveyor's chain of 22 yards contains 10 fathoms, 1 foot, 10.896 inches; 100 French fathoms contain 9.6916515 chains; the lineal arpent extends 63.9654 yards; the league of Louisiana, 244.2296 American chains; an acre is equal to 1.1829 superficial arpents of Louisiana; a square mile is equal to 756.1424 square arpens. One may calculate in general that the French foot is to the American as 16 is to 15. A square arpent is equal to 4,091.5724 square yards.

5th. It does not appear that the Spanish weights and measures have ever been used.

6th. The new superficial measures are very seldom used: the French and Spanish population will find it difficult to make use of them, as well as of the lineal measures.

7th. Neither Mr. Pilié, city surveyor, nor myself, know of any police law relative to weights or measures; and, although I have had an opportunity of reading a great deal, I have never fallen in with one of that nature. We know not of any French or Spanish standard in this country. The French and Spanish records of surveys (procés verbaux) are reported entirely according to the measures of Paris.

8th. The ancient inhabitants are well enough satisfied with the American weights, but all, and especially those in the country, find it very difficult to accustom themselves to the measures of length, and the superficial measures; and I think they will be long in doing so. This depends upon long contracted habits, and they cannot change these habits all at once. One must have gained great facility in calculation to be able to comprehend new superficial measures, at all times more difficult than measures of length or weights, especially when he sees no advantage to be derived from them.

The reports which I have presented of the ancient and new measures will be sufficient data for an accomptant to find out those others which I have not time to point out. Mr. Pilié and myself are very much pressed by business at present, and we beg Mr. Johnson to excuse us for not having done more, and in shorter time.

I beg him to be assured of the perfect consideration with which I am, &c.

BOUCHON,
Surveyor General of the state of Louisiana.
New Orleans, 9th October, 1820.

APPENDIX.

E 19—a.

INDIANA.

The Governor of Indiana to the Secretary of State.

CORYDON, *September* 9, 1817.

SIR: I have, herewith, the honor of submitting a report from the secretary of state for Indiana, relative to the regulation and standards for weights and measures in this state, in conformity to a resolution of the Senate of the United States of the 3d of March last.

Absence from the seat of government of our state prevented an earlier compliance with your request.

I have the honor to be,
With great respect,
Your obedient servant,
JONATHAN JENNINGS.

The Hon. the SECRETARY OF STATE.

E 19—b.

SECRETARY OF STATE'S OFFICE,

Corydon.

The Secretary of State, to whom was referred the resolution of the Senate of the United States of March 3d, 1817, requiring a statement relative to the regulations and standards for weights and measures in the several states, &c. has the honor to submit the following report, containing the regulations and standards for weights and measures as used, and now in force, in the state of Indiana.

One measure of one foot, or twelve inches English measure, so called; also, one measure of three feet, or thirty-six inches, English measure as aforesaid; also, one half bushel measure, for dry measure, which shall contain one thousand seventy-five and one-fifth solid inches; also, one gallon measure, which shall contain two hundred and thirty-one solid inches; which measures are to be of wood, or any metal the court may think proper; also, one set of weights, commouly called avoirdupois weight.

R. A. NEW, *Secretary*

*His Excellency the Governor
of the state of Indiana.*

APPENDIX.

E 20—a.

MISSISSIPPI.

The Governor of Mississippi to the Secretary of State.

NATCHEZ, *September* 17, 1818.

SIR: I have had the honor to receive your letter of the 8th ultimo. By an act of the legislature of the Mississippi territory, passed the 4th February, 1807, the treasurer thereof was directed to procure one set of weights and measures according to the standard of London, which was to serve as a general standard for the territory. He was also directed to procure a set for each county; the latter provision was never carried into effect. The legislature, therefore, by an act of the 23d December, 1815, authorized the treasurer to procure six sets of the same standard, and directed them to be deposited at certain public places, under the care of an officer, whose duties are prescribed by the act; this law was duly executed, and the weights and measures, which were procured, are now the standard by which weights and measures are regulated throughout the state. I enclose to you copies of the acts herein alluded to.

I have the honor to be,
With great respect,
Your obedient servant,
DAVID HOLMES.

To the Hon. JOHN Q. ADAMS.

E 20—b.

An Act relative to Weights and Measures, passed 4th February, 1807.

SEC. 1. The treasurer of the territory be, and he is hereby, authorized and required to procure, as soon as may be, at the public expense, one set of weights and measures, viz: one weight of fifty pounds, one of twenty-five, one of fourteen, two of six, two of four, two of two, and two of one pound avoirdupois weight, according to the standard of the United States, if one be established; but if there be none such, according to the standard of London; with proper scales for weights, together with measures of one of the length of one foot, and one of one yard, cloth measure; and the measures of one half bushel, one peck, and one half peck, dry measure; also, the measures of one gallon, one of half a gallon, one of one quart, one of one pint, one of half a pint, and one of one gill, wine measure, according to the above named standards; and the said weights, measures, and scale, shall be deposited with the said treasurer, to serve as a general standard for weights and measures within this territory, until otherwise directed by Congress.

SEC. 2. When the aforesaid weights, measures, and scales, shall be provided as aforesaid, the treasurer shall cause to be made or procured for each county within this territory, at the public expense, one set of weights, scales, and measures; and the last mentioned weights and measures shall be compared by the said treasurer with the aforesaid general standard, and, when found to agree therewith, shall be forthwith transmitted by him to the clerks of the several county courts in this territory.

SEC. 3. The said weights, measures, and scales, shall be kept by such person in each county as the county court shall direct, who shall take the following oath, viz: " I, A B, do solemnly swear, (or affirm, as the case may be,) that I will, in all things, act with justice and faithfulness in my appointment as keeper of the standard of weights and measures, for the county of ——, according to law, to the best of my skill and judgment. So help me God." And, immediately after such appointment, the clerks of the several county courts shall make known the same by advertisement, to be fixed at the door of the court house; and, also, by inserting the same in one of the public gazettes of this territory. And all persons, desirous of trying their weights and measures, may resort to the aforesaid county standard for that purpose; and, if they are found true, the person appointed to keep the said standard shall seal them with a seal, to be provided by the county court, at the expense of the county; and the persons appointed in the several counties, to keep the said county standard, shall be entitled to receive, for trying every steelyard, and giving certificate therefor, and for trying any other weights and measures, and sealing the same, fifty cents each, to be paid by the person for whom such service is rendered or done.

SEC. 4. Three months after the appointment of a person to keep the said county standard shall have been made as aforesaid, every person, or persons, who shall sell any commodity whatever by weight or measure that shall not correspond with the said county standard, or shall keep any such for the purpose of buying or selling with, shall, for every such offence, forfeit and pay the sum of ten dollars; recoverable, before a justice of the peace, by any person who will sue for the same, and applied to his own use.

SEC. 5. The territorial treasurer shall transmit to the clerks of the respective counties the weights, scales, and measures, as directed by the second section of this act, within twelve months from and after the passing hereof.

E 20—c.

An Act passed December 23, 1815.

SEC. 1. The treasurer of the territory be, and he is hereby, authorized and required to procure, as soon as may be, at the public expense, six sets of weights and measures as described in the act, entitled "An act establishing weights and measures in the Mississippi territory," passed February 4, 1807.

SEC. 2. When the sets of weights, measures, and seals, shall have been procured by the treasurer as authorized by this act, one set shall be conveyed, at the public expense, to Huntsville, in Madison county; one other set to some place on or near Pearl river; one other set to the town of St. Stephens; one other set to the town of Mobile; one other set to the town of Woodville; and the sixth set to the town of Port Gibson, and be placed in the hands of some person to be appointed by the governor; and the sets of weights and measures heretofore procured shall be forthwith delivered to some person residing in the city of Natchez, to be appointed in the manner above described. And the keepers of weights and measures, hereby authorized to be appointed, shall, previously to entering upon the discharge of their duties, take the oath or affirmation prescribed in the third section of the act above recited; and the said keepers shall, immediately after their appointment, make known the same by advertisement, to be fixed at the door of the court house of the counties in which they reside; and, also, by inserting the same in the nearest newspaper published in this territory. And, immediately after the appointment of the said keepers, all persons residing in any part of this territory, and desirous to try their weights and measures, may resort to any of the aforesaid standards for that purpose; and, if they are found true, the keeper of any of the said standards shall seal them with a seal, to be procured by the said treasurer at the expense of the territory; and the persons appointed keepers, as aforesaid, shall be entitled to receive the fees prescribed by the third section of the act above recited, for the services therein specified.

SEC. 3. Three months after public notice of the appointment of a person to keep the said standard in the city of Natchez, every person residing in any of the counties lying upon the Mississippi river, and that, three months after public notice of the appointment of the other keepers of weights and measures hereby authorized to be made, every person residing in this territory, who shall sell any commodity whatever by weight or measure, that shall not correspond with the standard hereby established, shall, for every such offence, forfeit and pay the sum of fifty dollars, recoverable before any justice of the quorum, or of the peace; to be paid to the county treasurer for county purposes.

SEC. 4. The sum of four hundred dollars, in addition to the eight hundred dollars heretofore appropriated for the purpose of procuring weights and measures, be, and the same is hereby, appropriated for the purposes specified.

APPENDIX.

E 21.

ILLINOIS.

An act regulating weights and measures.

SEC. 1. *Be it enacted by the People of the state of Illinois, represented in the General Assembly,* That it shall be the duty of the county commissioners in each and every county within this state, as soon a practicable after they are qualified to office, to procure, at the expense of their respective counties, one measure of one foot, or twelve inches, English measure so called; also, one measure of three feet, or thirty-six inches, English measure as aforesaid; also, one gallon liquid or wine measure, which shall contain two hundred and thirty-one cubic inches; one measure that shall contain one-fourth part, one measure that shall contain one-eighth part, one measure that shall contain one-sixteenth part, of the aforesaid liquid gallon, denominated quart, pint, and gill, each of which shall be made of some proper and durable metal; also, one half bushel measure for dry measure, which shall contain eighteen quarts, one pint, and one gill, of the above liquid or wine measure, the solid contents of which is equal to one thousand and seventy-five cubic inches, and fifty-nine hundredths of a cubic inch; likewise, one measure that shall contain one-fourth part of the aforesaid half bushel, or one gallon dry measure, which said half bushel and its fourth shall be made of copper or brass; also, a set of weights of one pound, one-half pound, one-fourth pound, one-eighth pound, and one-sixteenth pound, made of brass or iron; the integer of which shall be denominated one pound avoirdupois, and shall be equal in weight to seven thousand and twenty grains troy or gold weight; which measures and weights shall be kept by the clerk of the county commissioners for the purpose of trying and sealing the measures and weights used in their counties, for which purpose the said several clerks shall be provided with a suitable seal, or seals, with the name, or initials, of their respective counties inscribed thereon.

SEC. 2. *Be it further enacted,* That as soon as the county commissioners shall have furnished the measures and weights as aforesaid, they shall cause notice thereof to be given at the court house door, one month in succession immediately thereafter; and any person thereafter who shall knowingly buy, or sell, any commodity whatsoever, by measures or weights in their possession which shall not correspond with the county measures and weights, shall, for every such offence, being legally convicted thereof, forfeit and pay the sum of twenty dollars, for the use of the county where such offence shall have been committed, and costs of suit, to be recovered before any justice of the peace of said county. Every person desirous of having their measures and weights tried by the county standard shall apply to the clerk of the county commissioners, and if he find it correspond with the county standard, shall seal the same with the seal provided for that purpose; and said clerk is allowed to demand and receive such fees as now are, or hereafter may be, allowed by law.

This act to be in force from and after its passage.

APPENDIX.

E 22.

MISSOURI.

Extract from a law of the territory of Missouri, concerning weights and measures.

"Sec. 1. The several courts of common pleas [circuit courts] within this territory, shall provide for their respective counties, and at the expense of their said counties, one measure, of one foot, or twelve inches, English measure, so called: also, one measure of three feet, or thirty-six inches, English measure, as aforesaid, to be denominated one yard; also, one half bushel measure, which shall contain one thousand seventy five and one fifth solid inches, to be denominated dry measure: also, one gallon measure, which shall contain two hundred and thirty-one solid inches; one half gallon measure, which shall contain one hundred and fifteen and one half solid inches; and one quart measure, which shall contain fifty-seven and three fourths solid inches; which measures are to be of wood, or any metal the court shall think proper: also, one set of weights, commonly called avoirdupois weights; and one seal, with the initial of the county inscribed thereon; which measures, weights, and seal, shall be kept by the clerk of the court of common pleas [circuit court] in each county, for the purposes of trying and sealing the measures and weights used in their counties." L. M. T. July sess. 1813.

F.

PARIS, 16th July, 1817.

SIR: I had the honor to receive your letter dated London, the 5th of May last, together with the resolution of the Senate of the 3d of March, 1817, on the subject of weights and measures.

I accordingly enclose, in relation to those of France, *Tarbe's Manuel*, which is considered as the best elementary practical work on the subject; the Annuaire of the Board of Longitude, for the present year, in which you will find a very concise exposition of the principles of the system; the third volume of Delambre's *Base du Systeme metrique decimal*, which explains them at large; the *Connoissance des Tems*, for the year 1816, in which are found, pages 314 to 332, and particularly page 330, some subsequent observations on the pendulum; and some sheets of a journal now printing, which contains an additional note of De Prony, on the ratio of the metre to the English foot.

I have not sent the two first volumes of Delambre's work, which contain the details of the measurement of the meridian, from Dunkirk to Barcelona, as all the results are found in the third volume. The fourth volume, edited by Biot and Arrago, which, besides other matter, will give the measurement of the meridian from Barcelona to Formentera, its northern extension, to Greenwich, and through a part of Great Britain, and Biot's observations of the pendulum at several places, is not yet printed.

You requested that I might add such observations as might occur to me. The following are made, less on account of their intrinsic value, than because they may assist in explaining some points of the works enclosed.

The *legal metre*, or unit of the French linear measures, is presumed to be $\frac{1}{10000000}$ of the quarter of the meridian, from the equator to the pole, and has been deduced from the actual measurement of the arch from Formentera to Dunkirk, compared in order to calculate the flattening of the earth with the former measurement in Peru. This standard metre, such as it is definitively adopted by law, is equal to $443\frac{296}{1000}$ lignes (or 12th parts of an inch) of the old French measures. Doubts have however arisen whether it is truly the $\frac{1}{10000000}$ of the quarter of the meridian. Delambre, in his third volume, deducing the $\frac{1}{10000000}$ from the measurement of the arch from Barcelona to Dunkirk alone, had made it equal to $443\frac{322}{1000}$ lignes. The calculations of the flattening of the earth, as deduced solely from theory, would give a result nearer to the legal metre. The measurement made in England of an arch of the meridian, would seem to affect those conclusions. But it must be observed that an error of two seconds in the latitude observed, would, in the long arch measured from Formentera to Dunkirk, produce a difference in the length of the metre, equal to the difference between Delambre's calculation and the legal metre; that the same error would, in the much shorter arch measured in England, produce a much greater difference; and that the most candid astronomers acknowledge that, with the instruments now in use, such an error (of two seconds of latitude) is possible. Although, therefore, it seems probable that the legal metre is a little too short, and it seems to be regretted that it was not made equal to $443\frac{3}{10}$ lignes, it is, upon the whole, nearly as exact as was practicable in the present state of science. But, should more extensive measurements of the meridian, greater improvements in the instruments, and a more precise knowledge of all the elements which affect the observations, lead to a still more correct calculation of the true length of the quarter of the meridian, the standard, or legal French metre, would remain as it now is, the unit or basis of the system of French measures, and the only difference would be, that, instead of being, as now presumed, $\frac{1}{10000000}$, it might be found to be $\frac{1}{10000100}$, or some other not far distant fraction of the quarter of the meridian. And this difference would be less than the errors which will always take place in the confection, not of a standard metre executed with every possible care, but of the measures used for common purposes.

For the purpose of ascertaining the ratio of the length of the metre to that of the pendulum, observations have been made with great

APPENDIX.

care in several places; but those at Paris by Borda and Cassini, inserted in Delambre's third volume, are the only ones (with the exception of those mentioned in the *Connoissance des Tems*, page 314 to 322) which are given in detail by the French writers. Those of Biot and others will appear in Delambre's fourth volume: but their general result is given in the Connoissance des Tems, and it thereby appears that the length of the pendulum making 100,000 oscillations in 24 hours (in vacuo and at the freezing point of water) and referred to the level of the sea, is at Paris, in latitude 48° 50′ 14″, equal to $\frac{741993}{1000000}$ of the metre; at Bordeaux, in latitude 44° 50′ 25″, equal to $\frac{74161}{100000}$; and at Formentera, in latitude 33° 39′ 56″, equal to $\frac{74126}{100000}$. The final calculation of Borda for Paris, of the second pendulum (which makes 86,400 oscillations in 24 hours) is $\frac{993833}{1000000}$ of the metre, and $440\frac{560}{1000}$ lignes old French measure.

From the metre are immediately deduced, by a descending and ascending decimal ratio, all the French measures, linear, superficial, and of capacity, such as they are found every where, and which require no observation.

The unit of weight, which is in fact the kilogramme or 1000 grammes, has been determined by ascertaining the weight in vacuo of the $\frac{1}{1000}$ part of a cubic metre of distilled water at its maximum of density, which is 4° of the centigrade thermometer above the freezing point, corresponding to 39.6 of Fahrenheit. The experiments, which in every system of measures which may be adopted, are extremely important, were made with great care and skill by Lefevre Gineau; the substance is found in Delambre's third volume, but the detailed account appears to have been unfortunately lost. The old French pound, poids de marc, is $\frac{4895}{10000}$ of the kilogramme.

From the kilogramme or gramme has been deduced, by a decimal ratio, the whole system of French weights, including also that of moneys, in the manner stated in all the elementary works.

The ratio of the English foot to the old and new French measures, may have been lately ascertained in England, where a correct standard metre of platina has been sent. By de Prony's experiments the metre is equal to 39.3827 English inches, and the old French foot to 12.1232 English inches.

It is probable that the second pendulum will, at Washington, be found nearly equal to $\frac{993}{1000}$ of a metre, and to 39.107 English inches.

The great advantages of the French system seems to consist, 1st. In having an unit from which derives the whole by a decimal ratio; 2dly. In having ascertained the ratio of that unit to constant quantities, (the quarter of the meridian, and the length of the pendulum) so as to be able to perpetuate, and, in case of accident, to make new standard measures, perfectly similar to those now established by law: 3dly. In the great correctness of the experiments by which this ratio has been ascertained, and in the similar care bestowed on the confection of the standard measures and weights.

These advantages are, in a great degree, independent of the substitution of new to ancient measures, which has in practice met with such difficulties. Provided the ratio of the old French foot to the natural and constant measures of the earth and of the pendulum, had been ascertained, that foot might have been preserved, introducing only its decimal subdivision.

I will, for the present, only add, that as one of the first steps, if any plan is adopted on that subject in the United States, must be to ascertain the ratio of our foot to natural measures, it will be important to obtain the observations (not yet printed) of Biot, which, together with those of Borda, give all the information necessary for correct experiments of the length of the pendulum. As we will not probably very soon measure with sufficient correctness a considerable arch of the meridian, the best mode to obtain at once the ratio of our foot, both to the quarter of the meridian and to the French measure, will be to cause a metre of platina to be made here by Fortin. I could, I think, prevail on Mr. Arrago, or some other member of the board of longitude, to superintend the execution. Two have been completed for the Royal Society of London, and one is now on hand for the king of Prussia. The price will be about one hundred guineas. The brass metre made by order of Mr. Hassler, and which belongs to the United States, was executed by young Le Noir, and is not sufficiently correct for the purpose.

I am, &c.

ALBERT GALLATIN.

G.

Mr. Russell to the Secretary of State.

STOCKHOLM, 31st July, 1818.

SIR: The duplicate of your letter (No. 3.) of the 25th of May last, reached me on the 15th inst. but the original is not yet received. I have now the honor, in compliance with your request therein expressed, to communicate such information as I have been able to collect, " relative to the proceedings of this country for establishing uniformity in weights and measures," and which was required by the resolution of the Senate, to which you refer.

Not being made acquainted with the object of the Senate in framing that resolution, either by the resolution itself or by your letter, I have felt some uncertainty with regard to the precise import of the term " uniformity," as therein used. It may relate to the proceedings of any foreign country to establish, *within itself,* one sole standard for weights and measures respectively; or it may relate to the proceedings of several foreign countries to establish such a standard *in common,* or to the regulations of each particular country, for pre-

APPENDIX.

serving the constancy of its weights and measures, by an exact conformity to one or more standards, respectively, which may there exist, and thus, in regard to those respective standards, establishing *uniformity* throughout such country.

Were I to confine myself to either of these constructions, singly, I might err in my selection, and furnish information altogether inapplicable to the real object of the resolution. I have believed it safer, therefore, to communicate all the information which I have collected on the subject, considered in the three points of view above suggested. I shall even present some details, which, although obviously not called for by the resolution of the Senate, may be useful on some other occasion.

MEASURES.

Throughout Sweden there is but one measure, which was last established by the royal ordinance of King Frederick, in 1739. This measure appears, like many others now in use in Europe, to have been originally taken from the human foot and thumb. The Swedish foot is divided into twelve work inches (werktums) or decimally into ten inches. The work inches are used in building, handifraft, and commerce, and the decimal inches for geometrical mensuration.

Long measure.

To the foot thus divided into inches are all the Swedish measures of length, superficies, or capacity referrible :
2 Swedish feet = 1 Swedish (aln) ell ; 3 ells = 1 fathom (famm) ; 5 ells = 1 perch (stang) ; 36,000 Swedish feet = 18,000 Swedish ells = 1 Swedish mile.

Land measure.

56,000 Swedish square feet = 14,000 ells = 1 Swedish (tunnland) acre of land ; 1 tunnland = 2 spannlands = 4 half spannlands = 32 kapplands = 56 kannlands.

Dry measure.

1 Swedish cubic foot = 1,000 Swedish decimal cubic inches = 10 Swedish (kannor) cans ; 5,600 decimal cubic inches = 1 tunna (ton) = 2 span = 8 quarts (fjerdinger) = 32 kappar = 56 cans (kannor) ; 7 cans = 4 kappe.

Liquid measure.

48 cans (kannor) = 1 ton (tunna) = 2 half tons = 4 quarters (fjerdingar) = 8 eighths (attingar) ; 1 can (kann) = 2 stoopes (stoppar) = 4 half stoopes = 8 quarts (quarter) = 32 gills (ort or jungfrur.)

The can (kann) remains invariably the same in all measures, whether dry or liquid, in which it is used.

The measures of Sweden compared with those of France and England.

According to the official comparison made by the French philosophers Prony, Legendre, and Pictet, between the French standard metre and an English measure, made expressly by Troughton for that object, both these measures being at the temperature of freezing, the French metre is equal to 39.3827 English inches.

See Annals of Chemistry, for June, 1817, pa. 166. (Annales de Chimie et Physique.) Thus:

1 French metre = 39.2827 English inches (log. 5953056)
— 36.9413 old French inches (log. 5675125)
— 40.4175128 Swedish work inches (log. 6065695.)
— 33.6812606 Swedish dec. inches (log. 5273883.)

1,000 English feet = 1026.275 Swedish feet ⎫ the proportion between the decimal inches is the same. ⎫ (log. 0112639)
974.397 English feet = 1,000 Swedish feet. ⎭ ⎭ (log. 9887361)

1,000 English duodecimal inches = 1026.275 Swedish work inches.
1,000 do. do. do. = 855.229 Swedish decimal inches (log. 9320827.)
1169.276 English duodecimal inches = 1,000 Swedish decimal inches (log. 0679173.)
1,000 English sq. feet = 1053.24 Swedish sq. feet (log. 0225278.)
949.945 do. do. = 1,000 do. do. (log. 9774722.)
1,000 English duodecimal sq. inches = 731.417 Swedish decimal sq. inches (log. 8641654.)
1367.21 English duodecimal sq. inches = 1,000 Swedish decimal sq. inches (log. 1358346.)
1 Swedish acre (tunnland) = 56,000 Swedish sq. feet = 53,169 English sq. feet (log. 7256602.)
1,000 English cubic inches = 625.53 Swedish cubic decimal inches (log. 7962481.)
1 Swedish can (kann) = 100 Swedish cubic decimal inches = 159.864 English cubic inches (log. 2037519.)
1 Swedish meal ton (maltunna) = 5,600 Swedish decimal cubic inches = 8952.41 English cubic inches

This is divided into 8 fjerdingar, each 700 Swedish decimal cubic inches = 1119.048 English cubic inches.

In commerce there is also a corn ton (sparnmaltunna) = 6,300 Swedish decimal cubic inches = 10,071 English cubic inches; Which is in fact the same measure of exactly 5,600 Swed. cub. dec. in. But in the sale of all kinds of grain is added thereto 1 fjerding or eighth part, } 700

6,300

APPENDIX.

It was formerly the practice in the sale of grain to give *heaped* measure; but as this practice, for want of precision, occasioned continual disputes, it was at length abolished by law, and the addition of one-eighth part ordained in its stead; that is, 9 tons make 8 of corn, and the measure now is not to be *heaped* or *shaken*.

WEIGHTS.

Mintvigt, (mint weight.)

The Swedish mint weight, or that with which gold and silver are weighed at the mint and at the bank, when these metals are left for coining, is divided into the *mark, lod, qvintin,* and *ass :* and in respect to the *fineness* of silver, into the *mark, lod,* and *gran* (grain); and in respect to the fineness of gold, into the *mark, karat,* (carat) and *gran,* (grain.)

1 Mark, mint weight, = 16 lod = 64 qvintin = 4,384 (troyske) Dutch ass.

Medicenalsvigt, (apothecary weight.)

1 pound (libra) medicenalsvigt, is divided into 12 ounces, 1 ounce into 8 drams, 1 dram into 3 scruples, 1 scruple into 60 grains.

Thus, 1 libra = 12 ounces = 96 drams = 288 scruples = 5,760 grains = 7,416 (troyske) Dutch ass.

Victualievigt, (provision weight.)

The Swedish victualievigt is divided into *sheppund,* (shippounds) *centner* (hundreds,) *lispund,* and *marks,* or *skulpund.* Thus, 1 *shippound* = 4 *centner* = 5 *lispounds* = 400 *marks,* or *skulpounds.* The *skulpound* is divided into *lod* and *ass,* and 1 *skulpound* = 32 *lod* = 8,488 *ass.*

N. B. The centner is generally omitted in accounts, and one shippound divided at once into 20 lispounds.

Metalsvigt, Stapelstadsvigt, or Exportationsvigt.

The weight which is called by these three names is divided, like the victualievigt, into shippounds, centner, lispounds, skulpounds, lod, and ass, of the same relative value. The skulpound is also divided into fourth, eighth, and sixteenth parts.

Uppstadsvigt, Bergsvigt, and Tackjernsvigt.

These are three distinct weights, but are divided and subdivided in the same manner as the provision weight and the exportation weight.

APPENDIX.

Application of the several weights.

The use of the *mint weight* is already explained. The *medicenalsvigt* is, as the term imports, for weighing drugs and medicines. The *victualievigt* is that which is most generally and frequently used in Sweden. With it are weighed all kinds of provisions, and all merchandise which is sold within the country, or exported abroad by weight, excepting those articles only which specially appertain to the other sorts of weights herein mentioned. The *metalsvight*, stapelstadsvigt, or exportationsvigt, is applied exclusively to weighing iron, steel, copper, and other gross metals, for *exportation abroad*.

Bergsvigt, also called *Bergshammervigt*, (that is *weight, or mine hammerweight*,) is the weight used *at the forges*, for iron intended for *home consumption*, and to be sent into the interior, or to the *uppstads*, which are towns or cities whence no exportation abroad is allowed, there being at those places no custom houses for this purpose. But iron sent to the *stapelstads*, or cities whence exportation abroad is permitted, and at which there are custom houses for this purpose, is weighed *at the forges*, by the *metalls-stapelstads*, or *exportationsvigt*.

Uppstads weight is that used at *these places*, or any where in the *interior* where iron is sold for *home consumption*.

Tackjernsvigt (pig iron weight) is exclusively used throughout Sweden for weighing pig iron to the workmen, who are to forge it into bar iron for account of the proprietor.

Comparative view.

The Swedish *ducat* ought to consist of gold of the fineness of 23 carats and 5 grains, and to weigh gross $72\frac{18}{60}$ Dutch ass, or $62\frac{1}{2}$ ducats make one mark, mint weight, = 4384 Dutch ass (nearly.) One mark of gold of the fineness of 24 carats, gives $62\frac{2}{281}$ ducats. The addition is according to the alloy. The *remedium (mintage)* which is allowed for ducats, is one grain in the fineness, and one ass in the weight, each piece.

The Dutch ducat contains gold of the fineness of 23 carats and 8 grains, with a remedium of one grain per piece, of $72\frac{1}{2}$ Dutch ass. At least 72 pieces, new ducats, ought to weigh 5088 ass, or 159 angels.

Swedish silver coin, whole rix dollars, two thirds rix dollars, and one third rix dollars, consists of silver of the fineness of 14 lod one grain, with a remedium of one grain; one dollar of which ought to weigh gross $608\frac{4}{5}$ Dutch ass, or 36 rix dollars ought to weigh 5 marks mint weight:

1 *whole* dollar, of fine silver $534\frac{2}{3}$ ass.

1 *mark*, mint weight of fine silver, gives $8\frac{196094}{1000000}$ rix dollars.

1 *sixth* part of a rix dollar, or 8 styck, (shillings) half the fineness of silver, 11 lod 1 grain.

1 *twelfth* part of a rix dollar, or 4 styck do. 8 lod 2 grains.

1 *twenty-fourth* do. 2 styck. do. 6 lod 2 grains.

All with a remedium of one grain. All, both the whole dollar and the parts of the dollar, have each its full value in silver, and the copper alloy is thereto superadded.

Wrought silver ought to have the fineness of 13 lod and four grains, with a remedium of 2¼ grains. If the fineness be not thirteen lod, the vessel shall not on that account be broken up, but subjected to a double control duty, which makes it cost more than if the silver was of the requisite fineness.

All wrought silver is sold according to the lod victualievigt; and all wrought gold according to its weight in ducats.

The fineness of crown gold, so called, (kronguld) for different vessels, or other articles, is 18 carats. In victualievigt, as above, one skulpound = 8488 Dutch ass. The skulpound victualievigt is divided also into lod and quintin: thus, 1 skulpound = 32 lod, and 1 lod = 4 quintin = 276½ ass. One ounce, or 8 drams, *medecinalsvigt*, are equal to 9 quintin, *victualievigt*, or 1 *libra* = 7416 Dutch ass. *Metallsvigt* is $\frac{4}{5}$ of victualievigt. Thus, 16 skulpounds, victualievigt = 20 skulpounds metallsvigt; and one skulpound metallsvigt = $7078\frac{4}{10}$ Dutch ass.

Uppstads weight has a shippound = 421 skulpounds, metallsvigt, or 20 Uppstadsvigt, = $21\frac{1}{20}$ metalls or stapelstadsvigt. *Bergshammersvigt* has a shippound, = 442 skulpounds, *metallsvigt*, or 20 *bergshammersvigt*, = $22\frac{1}{10}$ *metalls*, or *stapelstadsvigt*.

The reason for this difference in the two last mentioned instances, is, that the iron being valued at the uppstads, and at the stapelstads, respectively, at the same price per shippound as at the forges, this factitious increase of weight has been devised to cover the expense of transportation to the respective places of delivery. Tackjernvigt has a shippound = 26 lispounds, or 520 skulpounds, bergshammersvigt, or 20 tackjernvigt,=26 bergshammersvigt, which means simply this, that the forgeman, for every 26 pounds of pig iron which he receives, must deliver to the proprietor 20 pounds of bar iron. From the *data* here furnished, all the weights of Sweden may be reduced to the victualievigt, or to the (troyske) Dutch ass. I will here observe that the word "*troyske*," which is used in the Swedish publications on weights and measures, applied to the ass, cannot be satisfactorily translated into other languages, as its derivative meaning is lost. Several learned men, however, have told me that they conjecture that it came from Holland. I have therefore translated it "Dutch."

To compare our weights with those of Sweden, the following *data* are believed to be sufficient.

1 skulpound, Swedish victualievigt, = 8848 Dutch ass.
1 pound English troy weight — 7766 do.
1 pound English avoirdupois weight = 9489 do.
1000 pounds English troy weight = 877.7124775 skulpounds, Swedish victualievigt, = 1000 pounds English avoirdupois weight, = 1066.794476 skulpounds, Swedish victualievigt, = 1333.49355 Swedish skulpounds, metalls, staplestads, or exportingvigt.

Uniformity.

With regard to uniformity, in the *first* sense above suggested, you will perceive that it exists in Sweden, in respect to *measures*, so far as the different uses of those measures will easily admit, as they are all reducible to the Swedish foot, lineal, square, or cubic, divided into 10 or 12 parts or inches. In respect to Swedish *weights,* such uniformity does not exist. The mint, apothecary, and provision weights, rest on foundations entirely distinct from each other, and the principal unit of each is divided into parts, with denominations and quantities peculiar to themselves. Although metalls, uppstads, bergs, and tackjerns weights, have indirectly a certain relation to provision weight, and the principal unit of each is divided into parts of the same names, and relative quantities, yet their diversity is sufficiently obvious from their different uses, and the different standards to which they must necessarily be made to conform. With regard to uniformity, in the *second* sense above suggested, there have been no proceedings whatever in this country for the purpose of establishing it: indeed the policy, as well as the habits, of the people appear to be opposed to its adoption. The only time when this question has been agitated here, was upon the receipt of the circular which the French government, soon after the introduction of the new weights and measures in France, addressed to the governments of other countries, recommending an universal conformity to the new standard. But there never, for a moment, existed here a disposition to accede to that proposition. All the convenience and facilities afforded by the plan proposed could not prevail against the prejudices opposed to it. The people consider their ancient customs as a constituent part of their rights, and would defend their old weights and measures as attributes of their liberty and independence.

The bankers object to the *decimal* system, because it does not readily admit of *thirds,* in which they pretend to have a great interest. And the politicians affect to believe that a diversity in weights and measures is necessary for the preservation of a diversity in governments, and that the adoption of general uniformity would facilitate universal conquest. I will not detain you with a comment on these objections.

With regard to uniformity in the *third* sense above suggested, which consists in preserving the *constancy* of weights and measures, by an exact conformity to one or more standards, respectively, I shall now present to you the regulations of Sweden.

Standard weights and measures are required by law to be kept at Stockholm, in the college of commerce and the mines, in the office of the receiver of the revenue, in the land surveyor's office, and in the City Hall. Standard weights and measures, adjusted by those of Stockholm, are also kept in the offices of the receiver of the revenue, and of the land surveyor in all the other towns or cities, and at all the parish churches in the country. The standard measures, the ell and the foot, are made of brass or steel. The standard weights are made of bronze. It is the duty of the land surveyors, in the interior, to take care that the weights and measures, in their respective districts, conform to the standards, and they are allowed a compensation for so doing.

The land surveyor at Stockholm is authorized and required to inspect the conduct of the land surveyors in the interior, and to see that they do their duty herein.

The measure for all kinds of grain, and for flour, salt, beef, and fish, must be made *square*, to promote the facility of ascertaining their justness, by the foot measure. But *round* measures are allowed for measuring salt, grain, and *other dry merchandise, on board of vessels when imported from foreign countries*. This round measure must be assayed once in every two months during the season of navigation, and while it is used a square measure ought to be at hand, with which to compare it. For measuring sea-coal, charcoal, chalk, and lime, the round measure is generally allowed. The seller is permitted to measure his own goods, but he must then pay the ordinary fee to the public measurer. Square measures are assayed by the foot rule, and round measures, if tight, by water, if not tight, by wheat or flaxseed. All measures for use ought to be made of seasoned wood, and their capacity stamped on the inside. To distinguish between the several sorts of weights, the victualievigt ought to be made round, and the metallsvigt *hexagonal*, and these weights, when used in the stapelstads, are to be marked with a *round* stamp, and, when used in the interior, with a square stamp. The bergshammersvigt, ought to be made triangular.

Steelyards were prohibited in 1638 and 1665, and were again prohibited by the ordinance of 1739, excepting for ore, and pig iron, by proprietors, for their own satisfaction; and also in such places where no regular weights are to be had. The ordinance of 1739, also excepts for the buying and selling of fish, and for domestic use, also for country people who bring their produce to market; in this case, however, the steelyards must be regularly adjusted, and not of a greater power than two lispounds, victualievigt. The large weights for use are made of iron or bronze, and adjusted with solder or lead, on which the stamp is impressed. The small weights are made of bronze; indeed, an ordinance of Gustavus the Third required *all* weights to be made of bronze, but this ordinance has had hitherto but a partial execution. Every person who uses weights or measures not conformable to the standards established by law is liable to a penalty.

I have now furnished all the information of any interest, which I have been able to collect here, relative to weights and measures, and I trust it will be found at least sufficient to satisfy the resolution of the Senate, so far as that resolution has any relation to Sweden.

I have the honor to be, with very great respect,
Sir,
Your faithful and obedient servant,
JONA. RUSSELL.

To the Hon. JOHN QUINCY ADAMS.

LIST OF PAPERS IN THE APPENDIX.

A. 1. Weights and measures used in the several custom houses of the United States.
    2. Table of ditto.
    3. Note on the weights and measures used in the District of Columbia.

B. 1. Table of the several English statute measures of capacity at different periods, compared with existing standards.
    2. Notes on ditto.

C.   Note on the proportional value of the pound sterling and the dollar.

D. 1. Mr. F. R. Hassler to the Secretary of State, Oct. 16, 1819.
    2. Mr. Hassler's comparison of French and English standard measures of length.

E.   Weights and measures of the several states, viz:

E. 1. New Hampshire.
    a. Governor Plumer to the Secretary of State, Nov. 21, 1818.
    b. Law of May 13, 1718.
    c.   do.   6 Geo. III.
    d.   do.   Dec. 15, 1797.
    e. Governor Plumer to the Secretary of State, Jan. 4, 1819.

E. 2. Vermont.
    a. Governor Galusha to the Secretary of State, Jan. 20, 1818.
    b. Law.

E. 3. Massachusetts.
    a. Governor Brooks to the Secretary of State, Oct. 24, 1817.
    b. Secretary Bradford to Gov. Brooks, Sept. 5, 1817.

E. 4. Rhode Island.
    Governor Knight to the Secretary of State, Sept. 5, 1817.

E. 5. Connecticut.
    a. Governor Wolcott to the Secretary of State, Aug. 5, 1817.
    b. Secretary Day to the Secretary of State, Aug. 12, 1817.
    c. Law of October session, 1800.
    d.   do.   do.   1801.
    e.   do.   May session, 1810.

244   LIST OF PAPERS.

E. 6. New York.   a. Governor Clinton to the Secretary of
                          State, Sept. 4, 1817.
                       b. Law of June 19, 1703.
                       c. do.   April 10, 1784.
                       d. do.   March 7, 1788. (Ext.)
                       e. do.   Feb. 2, 1804.
                       f. do.   March 24, 1809.
                       g. do.   March 19, 1813.
E. 7. New Jersey.   Governor Williamson to the Secretary
                          of State, Sept. 20, 1817.
E. 8. Pennsylvania.   a. Secretary Boileau to the Secretary of
                          State, Aug. 26, 1817.
                       b. Mr. J. Meer to Mr. Boileau, Aug. 20,
                          1817.
E. 9. Delaware.   a. Secretary Ridgely to the Secretary of
                          State, Nov. 7, 1818.
                       b. Law of 1704.
                       c. Mr. I. Booth to Mr. Ridgely, Oct. 24,
                          1818.
E. 10. Maryland.   a. Mr. Pinkney to the Secretary of State,
                          Aug. 9, 1817.
                       b. Governor Goldsborough to the Secre-
                          tary of State, Dec. 1. 1819.
                       c. Law of April, 1715.
                       d. do.   Dec. 20, 1765.
E. 11. Virginia.   a. Governor Preston to the Secretary of
                          State, Aug. 15, 1818.
                       b. Law of August, 1734.
                       c. do.   Dec. 26, 1792.
E. 12. North Carolina. a. Governor Miller to the Secretary of
                          State, Aug. 19, 1817.
                       b. Law.
E. 13. South Carolina. a. Governor Pickens to the Secretary of
                          State, Sept. 10, 1818.
                       b. Law of April 12, 1768.
                       c. do.   March 17, 1785.
E. 14. Georgia.   a. Governor Clark to the Secretary of
                          State, Dec. 5, 1819.
                       b. Law of Dec. 10, 1803.
                       c. Ordinance of the city council of Au-
                          gusta. (Ext.)
E. 15. Kentucky.   a. Lieutenant Governor Slaughter to the
                          Secretary of State, Nov. 26, 1817.
                       b. Law of Dec. 11, 1798.
E. 16. Tennessee.   Governor McMinn to the Secretary of
                          State, Nov. 24, 1819.
E. 17. Ohio.   a. Governor Worthington to the Secreta-
                          ry of State, Aug. 18, 1817.

E. 17. Ohio.
                    b. Secretary McLean to the Secretary of State, Sept. 13, 1817.
                    c. Law of Jan. 22, 1811. (Ext.)

E. 18. Louisiana.
                    a. Governor Villeré to the Secretary of State, Sept. 15, 1817.
                    b. Law of Dec. 21, 1814.
                    c. M. Bouchon's statement of weights and Measures in Louisiana before its cession. (Translation.)

E. 19. Indiana.
                    a. Governor Jennings to the Secretary of State, Sept. 9, 1817.
                    b. Secretary New's report.

E. 20. Mississippi.
                    a. Governor Holmes to the Secretary of State, Sept. 17, 1818.
                    b. Law of Feb. 4, 1807.
                    c.   do.   Dec. 23, 1815.

E. 21. Illinois.   do.   March 22, 1819.
E. 22. Missouri.   do.   July, 1813.

F. Mr. A. Gallatin to the Secretary of State, July 16, 1817, on the weights and measures of France.

G. Mr. Jonathan Russell to the Secretary of State, July 31, 1818, on the weights and measures of Sweden.

CPSIA information can be obtained
at www.ICGtesting.com
Printed in the USA
LVOW13s0034201017
553104LV00014B/488/P